# The politics of identity and civil society in Britain and Germany

MANCHESTER
1824

Manchester University Press

## Critical Labour Movement Studies

Series editors
**John Callaghan**
**Steven Fielding**
**Steve Ludlam**

Already published in the series

Jenny Andersson, *Between growth and security: Swedish social democracy from a strong society to a third way*

John Callaghan, Steven Fielding and Steve Ludlam (eds), *Interpreting the Labour Party: approaches to Labour politics and history*

Andrew Gamble, Steve Ludlam, Andrew Taylor and Stephen Wood (eds), *Labour, the state, social movements and the challenge of neo-liberal globalism*

Dianne Hayter, *Fightback! Labour's traditional right in the 1970s and 1980s*

Jonas Hinnfors, *Reinterpreting social democracy: a history of stability in the British Labour Party and Swedish Social Democratic Party*

Declan McHugh, *Labour in the city: the development of the Labour Party in Manchester, 1918–31*

Jeremy Nuttall, *Psychological socialism: the Labour Party and qualities of mind and character, 1931 to the present*

Ben Jackson, *Equality and the British Left: A study in progressive political thought, 1900–64*

Lucy Robinson, *Gay men and the left in post-war Britain: How the personal got political*

# The politics of identity and civil society in Britain and Germany

## Miners in the Ruhr and south Wales 1890–1926

Leighton S. James

**Manchester University Press**
Manchester and New York
distributed exclusively in the USA by Palgrave

*Published by* Manchester University Press
Oxford Road, Manchester M13 9NR, UK
*and* Room 400, 175 Fifth Avenue, New York, NY 10010, USA
www.manchesteruniversitypress.co.uk

*Distributed exclusively in the USA by*
Palgrave, 175 Fifth Avenue, New York,
NY 10010, USA

*Distributed exclusively in Canada by*
UBC Press, University of British Columbia, 2029 West Mall,
Vancouver, BC, Canada V6T 1Z2

*British Library Cataloguing-in-Publication Data*
A catalogue record for this book is available from the British Library

*Library of Congress Cataloging-in-Publication Data applied for*

ISBN 978 0 7190 7497 4 *hardback*

First published 2008

17 16 15 14 13 12 11 10 09 08 10 9 8 7 6 5 4 3 2 1

Typeset
by Action Publishing Technology Ltd, Gloucester
Printed and bound in Great Britain
by Antony Rowe Ltd, Chippenham, Wiltshire

# Contents

# Series editors' foreword

The start of the twenty-first century is superficially an inauspicious time to study labour movements. Political parties once associated with the working class have seemingly embraced capitalism. The trade unions with which these parties were once linked have suffered near-fatal reverses. The industrial proletariat looks both divided and in rapid decline. The development of multi-level governance, prompted by 'globalisation' has furthermore apparently destroyed the institutional context for advancing the labour 'interest'. Many consequently now look on terms such as the 'working class', 'socialism' and 'the labour movement' as politically and historically redundant.

The purpose of this series is to give a platform to those students of labour movements who challenge, or develop, established ways of thinking and so demonstrate the continued vitality of the subject and the work of those interested in it. For despite appearances, many social democratic parties remain important competitors for national office and proffer distinctive programmes. Unions still impede the free flow of 'market forces'. If workers are a more diverse body and have exchanged blue collars for white, insecurity remains an everyday problem. The new institutional and global context is moreover as much of an opportunity as a threat. Yet, it cannot be doubted that, compared with the immediate post-1945 period, at the beginning of the new millennium, what many still refer to as the 'labour movement' is much less influential. Whether this should be considered a time of retreat or reconfiguration is unclear – and a question the series aims to clarify

The series will not only give a voice to studies of particular national bodies but will also promote comparative works that contrast experiences across time and geography. This entails taking due account of the political, economic and cultural settings in which labour movements have operated. In particular this involves taking the past seriously as a way of understanding the present as well as utilising sympathetic approaches drawn from sociology, economics and elsewhere.

John Callaghan
Steven Fielding
Steve Ludlam

# Tables

# Abbreviations

**Abbreviations used in the text**

| | |
|---|---|
| AAM | Amalgamated Association of Miners |
| AAU | Allgemeine Arbeiter Union (General Workers' Union) |
| ADAV | Allgemeiner Deutscher Arbeiterverein (General German Workers' Association) |
| AfSG | Archiv für Sozialgeschichte |
| CdI | Centralverbandes deutscher Industrieller (Central Association of German Industry) |
| CLC | Central Labour College |
| CMOC | Coal Mining Organisation Committee |
| CPGB | Communist Party of Great Britain |
| FAUD | Freie Arbeiter-Union Deutschland (Free Workers' Union, Germany) |
| FAUG | Freie Arbeiter-Union Gelsenkirchen (Free Workers' Union, Gelsenkirchen) |
| FCC | Food Control Committee |
| FUB | Freie Vereinigung der Bergarbeiter (Free Association of Miners) |
| FVdG | Freie Vereinigung deutscher Gewerkschaften (Free Association of German Trade Unions) |
| ILP | Independent Labour Party |
| KPD | Kommunistiche Partei Deutschlands (Communist Party of Germany) |
| MFGB | Mineworkers' Federation of Great Britain |
| MICUM | Mission Interalliée de Contrôle des Usines et Mines |
| MSPD | Mehrheits Sozialdemokratische Partei Deutschlands (Majority German Social Democratic Party) |
| MSWCOA | Monmouthshire and South Wales Coalowners' Association |
| RGO | Revolutionäre Gewerkschafts-Opposition Revolutionary Trade Union Opposition) |
| RILU | Red International of Labour Unions |
| RWKS | Rheinisch-Westfälische Kohlensyndikat |

| | |
|---|---|
| SDF | Social Democratic Federation |
| SPD | Sozialdemokratische Partei Deutschlands (Social Democratic Party of Germany) |
| SWMF | South Wales Miners' Federation |
| SWMIU | South Wales Miners' Industrial Union |
| TLC | Trades and Labour Council |
| UMA | United Mineworkers of America |
| URC | Unofficial Reform Committee |
| USPD | Unabhängige Sozialdemokratische Partei Deutschlands (Independent Social Democratic Party of Germany) |
| ZAG | Zentralarbeitsgemeinschaft (Central Works Community) |
| ZZP | Zjednoczenie Zawodowe Polskie (Polish Trade Union) |

**Abbreviations used in the notes**

| | |
|---|---|
| *AfSG* | Archiv für Sozialgeschichte |
| *AJPS* | American Journal of Political Science |
| *AL* | Aberdare Leader |
| *AVCECN* | Amman Valley Chronicle and East Carmarthen News |
| *BBA* | Bergbau Museum, Bochum |
| *Bk* | Bergknappe |
| BüK | Büro Kolpin |
| *BZ* | Bergarbeiter Zeitung |
| CCRO | Carmarthen County Records Office |
| *CEH* | Central European History |
| *CG* | Colliery Guardian |
| *CJ* | Carmarthen Journal |
| *CNWWGA* | The Celtic News and West Wales General Advertiser |
| *CWM* | Colliery Workers' Magazine |
| *DA* | Deutsche Arbeit |
| *DZ* | Dortmunder Zeitung |
| *EAZ* | Essener Arbeiter Zeitung |
| *EHR* | Economic History Review |
| *EVZ* | Essener Volkszeitung |
| *FPM* | Free Press of Monmouthshire |
| *GAZ* | Gelsenkirchener Allgemeine Zeitung |
| *GCT* | Glamorgan County Times |
| *GFPRL* | Glamorgan Free Press and Rhondda Leader |
| *GG* | The Glamorgan Gazette: Bridgend and Neath Chronicle and Central Glamorgan Gazette, Incorporated |
| *GH* | German History |
| GRO | Glamorgan Records Office |
| *HZ* | Historische Zeitschrift |
| IfSB | Institut für Soziale Bewegungen |
| *IRSH* | International Review of Social History |
| *IWK* | Internationale Wissenschaftliche Korrespondenz |

| | |
|---|---|
| IZF | Institut für Zeitungsforschung |
| *JCH* | Journal of Contemporary History |
| *JMH* | Journal of Modern History |
| LA | Landrat |
| *LV* | The Labour Voice |
| *ME* | Merthyr Express |
| *MEP* | Monmouthshire Evening Post |
| *MMSWA* | Monmouthshire Merlin and South Wales Advertiser |
| *MS* | Märkischer Sprecher |
| NLW | National Library of Wales, Aberystwyth |
| *NGMGH* | Neath Gazette and Mid-Glamorgan Herald |
| OBA | Oberbergamt |
| *PC* | Pontypridd Chronicle |
| *PG* | Porth Gazette |
| *PO* | Pontypridd Observer |
| PRO | Public Records Office |
| Reg. | Regierungspräsident |
| *RD* | Red Dawn |
| *RL* | The Rhondda Leader |
| *RS* | Rhondda Socialist |
| *RWAZ* | Rheinisch-Westfälische Arbeiter Zeitung |
| *RWZ* | Rheinisch-Westfälische Zeitung |
| *SH* | Social History |
| *SJPE* | Scottish Journal of Political Economy |
| STAB | Stadtarchiv Bochum |
| STAD | Staatsarchiv Düsseldorf |
| STAM | Staatsarchiv Münster |
| SWCA | South Wales Coalfield Archive |
| *SWDN* | South Wales Daily News |
| *SWDS* | South Wales Daily Star |
| *SWDT* | South Wales Daily Telegraph |
| *SWML* | South Wales Miners' Library |
| *SWA* | South Wales Argus |
| *SWW* | South Wales Worker |
| *T* | Tremonia |
| *TG* | Tarian Y Gweithiwr |
| TUC | Trade Union Library |
| UDC | Urban District Council |
| *Vb* | Volksblatt |
| *W* | The Welshman |
| *WHR* | Welsh History Review |
| *WK* | Westfälischer Kämpfer |
| *WM* | Western Mail |
| *WO* | Welsh Outlook |
| *WVZ* | Westfälische Volkszeitung |

The Ruhr industrial district

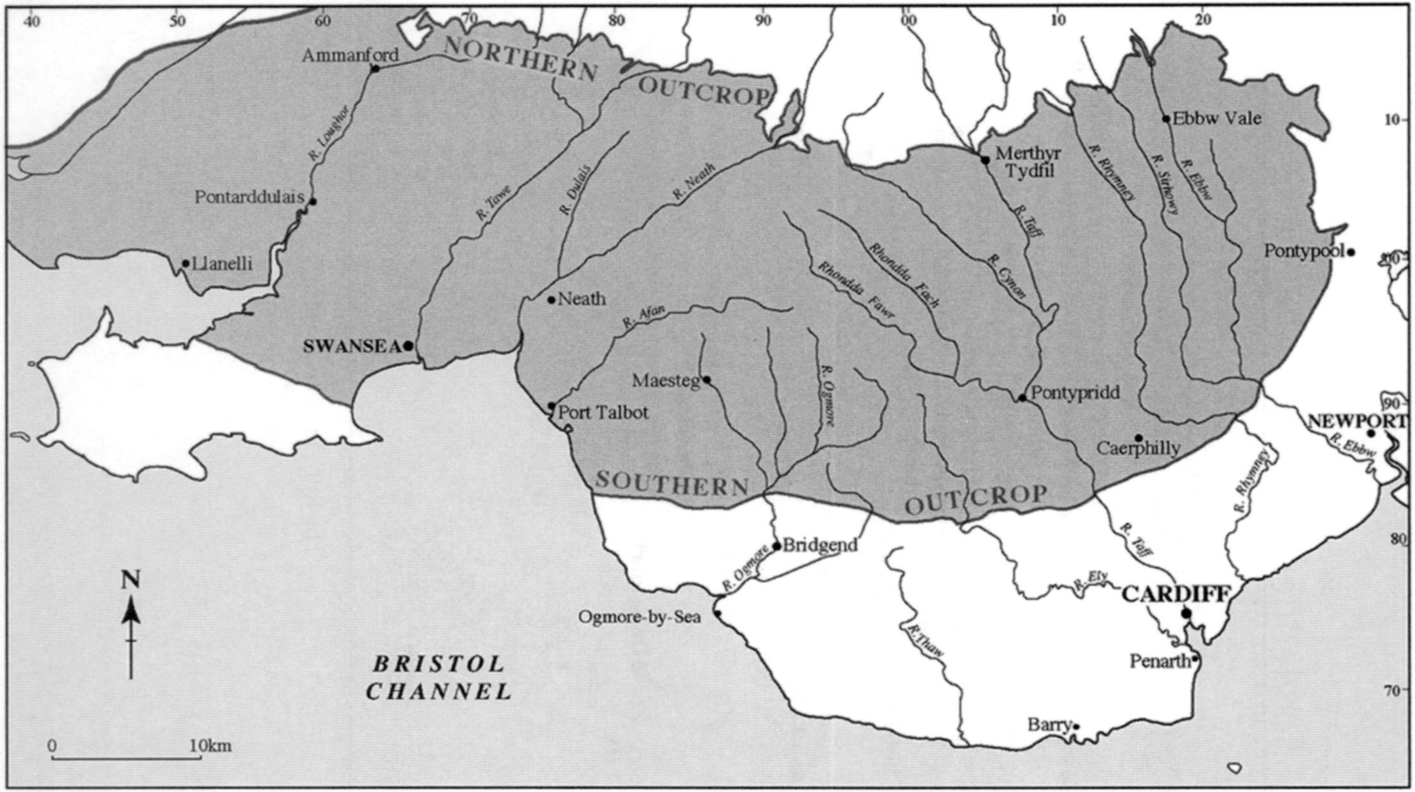


The south Wales coalfield

# Acknowledgements

I have accrued many debts while writing this book. First and foremost, I must thank Stefan Berger and Chris Williams. Both have provided invaluable constructive criticism and support throughout. Many others were also generous with their advice and encouragement, among them David Howell, Nina Fishman, Gareth Pritchard and Quentin Outram.

The Institut für Soziale Bewegungen provided the base for my research in Germany. I would like to thank Klaus Tenfelde, Peter Friedemann, Benjamin Ziemann, Holger Heith, and Kai Rawe for sharing their knowledge on German archives and providing insights on my topic generally. Brian McCook was kind enough to advise me on the role of the Polish community in the Ruhr and listen to my ideas over a glass of beer in the *Bermudadreieck*. The Deutscher Akademischer Austauch Dienst (German Academic Exchange Progamme) provided funds for my second research trip to Germany in 2002/3.

I owe a huge debt to many archivists and librarians in two countries. Diane White and the staff of the Goethe Institut Düsseldorf deserve a mention for they succeeded in teaching a reluctant mind some new tricks. Linda Norton at the National Museum at Cardiff was kind enough to draw the map of the Ruhr. Thanks go to Chris Williams for letting me reproduce the map of south Wales.

Finally, my family and Katherine Shelton have provided unstinting support. Without them this book would never have seen the light of day.

For my grandfather, William Edward Watkins

# Introduction

> [We] asked why the owners did not recognise the unions as they had done in Great Britain. [Our guide] replied that they did not do so because the unions were not united, and did not represent even half of the workmen employed; under these circumstances the workers had no guarantee that any agreement they entered into would be carried out by the men's side. As every Trade Unionist knows, the recognition of a union by the masters depends not so much on the goodwill of the latter as the power of the Trade Union. The miners' unions of Germany being split up on four parts have not the power to enforce this recognition, and the miners are, therefore, obliged to put up with many indignities which a strong union would not suffer for a moment. The case of the disunited German miners should be a warning to those persons in Great Britain who attempt from time to time to sow the spirit of disunion amongst our men.[1]

Such was the warning issued by Robert Smillie, President of the Miners' Federation of Great Britain, and Alfred Onions, Secretary of South Wales Miners' Federation (SWMF), following their visit to the Ruhr coalfield in 1910. In the same year the mining foreman and aspiring trade union leader Georg Werner penned a pamphlet following a visit to Britain entitled, *What can the German mining industry learn from England?*.[2] The British and German miners' movements were acutely aware of each other, their understanding shaped in the pre-war period through the regular International Miners' Congresses. Indeed, as these two examples show, miners' leaders engaged in comparison long before historians of the industry. Yet although Smillie, Onions and Werner were agreed that unity was the key to recognition, none of them provided a detailed explanation as to why the British unions had been able to achieve this state, while the German unions had not.

This book attempts to answer that question through a regional case study of the south Wales and the Ruhr coalfields. In the period between 1890 and 1926, south Wales exhibited a working-class political development, whereby a single union, the SWMF, and one party, the Labour Party, were able gradually to overcome sectionalist loyalties, dominate worker politics and enable a broad-based working-class identity to emerge. In the Ruhr region, by contrast, unions and parties representing working-class interests became increasingly fragmented. While all worker unions and political parties present in this region shared certain general beliefs regarding the dignity and rights of workers, the sectionalist worker identities of competing socialist, Christian and ethnic organisations prevented one dominant concept of working-class identity, and thus one governing form of working-class politics, from emerging. The book begins with the hypothesis that, on a comparative level, south Wales and the Ruhr were characterised during this period by divergent trends; the first by increasing homogeneity, the second by increasing fragmentation.

Unity and division were, of course, themes that occupied the whole of the British and German labour movements. Although this book presents a case study of two coalfields it feeds into a much wider literature on the miners' place in the respective histories of the German and the British proletariat. This question has exercised labour historians for decades. Early historians, close to the labour movement and their subjects, presented them as the vanguard of labour politics and trade union organisation.[3] This view, as will become evident in the following pages, paralleled the attitudes of the miners' leaders themselves. After all, they preferred to regard themselves as the pioneers and leading lights of national labour movements rather than leaden-footed stragglers. Even where trade unionism was a late development, as in south Wales, activists quickly came to believe that they had pushed their way to the forefront. Arthur Horner might have been concerned with the structure of the Miners' Federation of Great Britain (MFGB) when he advocated a national British miners' union in 1926, but he undoubtedly had one eye on the wider trade union movement when he claimed that 'existing trade unions belong to a past era'.[4]

Concepts such as the archetypal proletariat and radical miner implicitly reaffirmed the image of the miners as the most active of the working class. This image remained tenacious, and emotive given the miners' strikes of the 1970s and early 1980s.[5] But from the vantage point of the post-industrial twenty-first century this heroic narrative appears far too neat. Since the 1980s historians have revealed the complex and often fraught relationships among the miners themselves, let alone with other industrial workers. Both the British and German miners could be as inclined to conservatism as anyone else. Consequently, the radical miner has lost his explanatory, if not his romantic, appeal.[6] Yet the German and British miners have remained a fascination for labour historians, because, radical or not, they were major players in national labour movements. In politics they formed a sizable, if not united, voting block and an important source of aspiring politicians. The unions may not have been as radical or as representative of the working class as their leaders liked to think, but their influence is beyond doubt. In the nineteenth and early twentieth centuries the actions of the miners' unions and events in the coalfields attracted the notice of Governments, the press, and the public. And among those clamouring hardest for attention were south Wales and the Ruhr.

## Perspectives on south Wales and the Ruhr

South Wales and the Ruhr, as two of the most significant industrial regions in Britain and Germany respectively, have attracted the attention of many labour and social historians. Yet the historiographical traditions of the two regions are very different.[7] Much recent work on south Wales has been characterised by a microhistorical approach. The works of, amongst others, Chris Williams and Mike Lieven have been useful in deconstructing the metanarratives of previous historians such as Robin Page Arnot, Hywel Francis and Dai Smith. The latter presented the south Wales miners' movement as so 'intimately associated with its society' that to write separate histories of the union and the mining communities could be deemed almost a false dichotomy.[8] These histories concentrated upon the heroism of the miners' struggles and the search for 'alternative society'. More recently historians have revealed the complexity of local relations and questioned the previous notions of societal and organisational homogeneity.[9] Other historians of south Wales have moved away from the traditional areas of unionism and party politics to examine the role of women in south Wales and popular culture.[10]

The historiography of the Ruhr miners is comparatively longer and more complicated. One trend identified the miners as the radical force within the German labour movement after 1918.[11] However, this radical image never dominated and many historians questioned the miners' involvement in radical protest after the war. For example, the unrest and violence in the Ruhr in 1919 and 1920 was recently placed in the context of the material deprivation the miners experienced during the war and the following period of economic instability.[12] Other accounts have also concentrated on the miners' lack of solidarity and militancy. Klaus Tenfelde argues in his seminal work on the miners that changes in working practices in the last quarter of the nineteenth century led to a de-skilling of the workforce and a subsequent loss of autonomy and solidarity.[13] Finally, Stephen Hickey's structural account of the trade unionism and politics of the Ruhr miners in Wilhelmine Germany concentrates on the disunity and sets out the various obstacles to the creation of a unified labour movement.[14]

Such accounts, however, are faced with the difficulty of explaining the great strike waves that did occur in the Ruhr in 1889, 1905, 1917 and 1919. In response to this question, David Crew has argued in his study of Bochum that an occupational community was created among the miners. Individual miners carried this sense of community as they moved from mine to mine and it underpinned the episodic incidences of industrial action.[15]

Other German historians, influenced by the *Alltagsgeschichte* (the history of everyday life) approach, have turned to the study of working-class culture in the Ruhr.[16] Franz Josef Brüggemeier criticises Crew for neglecting to examine the nature of mining as a job and argues it was an important factor in the creation of solidarity.[17] Co-operation underground coupled with the miners' pride in their physically demanding world of work provided the basis for collective action.[18] Beyond the workplace the miners formed improvised networks as a means of self-help. Two manifestations of this informal support structure were the system of lodgers (*Schlafgänger*) and drinking clubs (*Schnappskasinos*).[19] These networks provided the necessary solidarity to support the various strike waves between 1889 and 1919, while the trade unions and Social Democratic Party of Germany (SPD), through their disapproval of these activities, imposed themselves between this informal network and the miners, thereby hampering expressions of solidarity.[20] *Alltagsgeschichte* shares many features with the older British social history.[21] *Alltagsgeschichte*'s concern with everyday culture, experiences and oral history have become well established within Welsh historiography, although without the debate that has accompanied its growth in Germany.

While there are some similarities in the social histories of south Wales and the Ruhr, the political historiographies of the two are characterised by very different concerns. The political history of nineteenth-century Wales is largely orientated to explaining the Liberal dominance of the principality and the emergence of a distinctive Welsh politics. Within this narrative the thriving industrial communities of south Wales played a crucial, although ultimately problematic role:[22] problematic because much of the work on south Wales's politics for the period between 1890 and the 1920s has concentrated on the 'rise of Labour/decline of Liberalism' paradigm. Like historians elsewhere in Britain, those of south Wales have largely been concerned with explaining this transition from Liberal to Labour. Various explanations have been advanced to explain the strange death of Liberal Britain. These have ranged from the exogenous (the Great War), to the domestic (social changes in British society) to the high political (schism in the Liberal Party). All these factors are undoubtedly significant, although recent microstudies have tended to point to the importance of local factors.[23] However, there remains an attachment to the distinctiveness of Welsh politics, in which the Labour Party inherited the radicalism of the Liberals that first brought Welsh issues to prominence.[24]

Historians of the Ruhr, on the other hand, have focused on explaining the structural basis of an essentially three-party system in the area before the Great War. In seeking to explain the particular political formations in the Ruhr the concept of milieu has often been invoked. British historians are largely unfamiliar with the term. It links political parties to distinct groups based on ethnic, religious or political identities. The milieux were made up of dense networks of organisations, including political parties and trade unions, and were consolidated by discriminatory state action. In this way the Catholic minority in Germany was forged into a milieu by Bismarck's laws against the Catholic Church in the 1870s.[25] Rainer Lepsius has argued that the stability of the Wilhelmine electoral system was conditioned by the existence of these sociocultural milieux. The parties were the overt political expressions of these various milieux and voting became a ritualised exercise in which each group used its ancillary organisations to turn out their supporters.[26]

Applying these ideas to the coalfield, historians of the Ruhr have identified four distinct types of milieux: Catholic, social-democratic, communist and national. These could also be further sub-divided into those based upon occupational subgroups, ethnic groups and organisations.[27] Karl Rohe has refined these distinctions and suggests that the Centre Party represented the model of the milieu-type party, while the Social Democrats, despite their numerous sporting and cultural associations, represented more of a *Sammlungsbewegung*, or protest movement. The entrenched nature of the Catholic milieu in the Ruhr meant that the SPD was at an electoral disadvantage, despite the predominantly working-class character of the region.[28]

This book departs from these existing historiographies in three ways. First, it breaks with the trend towards microstudies and returns to a coalfield-wide view of the history of miners' movements. While microhistorical accounts quite rightly question the old certainties that surround previous works and provide a deeper understanding of the complexity of mining communities, their usefulness for cross-regional comparisons that seek a more comprehensive theoretical framework to explain the development of such movements are more limited. Since the rationale of microstudies is to emphasise the local and specific, questions necessarily surround the typicality of the units of study. This may disturb some practitioners of microhistory, but a cross-regional comparison requires a regional perspective for both its subjects. While the caveats and the qualifications raised by local studies should be respected, to attempt to account for every difference between every community would risk reducing any work to an atomised account of local peculiarities.

The region itself is a problematic analytic category. Much effort and ink has been expended attempting to define the 'region'.[29] This work accepts that south Wales and the Ruhr can be seen as regions precisely because the protagonists in this study considered them so. This, however, raises other questions. How typical are south Wales and the Ruhr coalfields of other heavy industrial regions? Where do they fit within the British and German national context? In many respects both coalfields were atypical of other British and German industrial areas. Trade union leaders and politicians in the Ruhr, for example, were very conscious of its position as Germany's leading industrial area. South Wales, on the other hand, was but one coalfield among many. Yet, as will become evident later, south Wales unionists also believed the coalfield to be distinct from others. This was partly due to its orientation towards the export trade, but was also linked to notions of Welsh national identity. These caveats are not meant to suggest that the two coalfields were somehow outside the nation or untouched by national movements. Both the trade unions and the political parties were intimately involved with wider national and international movements. Union activists and politicians regularly met and interacted with counterparts from other regions and nations and were called upon to formulate responses to national and international events. As will be seen later, the south Wales miners were among the last to join the MFGB, but by the 1920s were one of the most radical unions in the federation. Meanwhile, the *Gewerkverein* served as a model for Christian trade unions in other industrial sectors. The industrial and economic importance of the Ruhr attracted the interest not only of successive governments but also a wide spectrum of political parties. Indeed, the case studies presented here are intended to be suggestive of developments elsewhere. Thus, these caveats are meant merely to indicate that many characteristics of the south Wales and Ruhr miners' movements were the result of specific regional factors that did not necessarily pertain outside the coalfields.

Secondly, the book adopts an explicitly comparative approach. Comparative history has its difficulties. Similar sources are not always to be found. For example, many trade union records have not survived in the Ruhr. On a theoretical level, there are difficulties in integrating two very different historiographical traditions. Specifically, the concept of the milieu, so important in the historiography of the Ruhr, has not been applied to south Wales. Indeed, it seems inapplicable since south Wales lacks the state pressures necessary to create such milieux in the late nineteenth century. Yet the significance of the concept in previous studies of the Ruhr means that it cannot be abandoned lightly.

Comparative history does, however, have the advantage of allowing the historian to test theories developed within one national or regional context. Several historians have already placed south Wales within a comparative perspective. David Gilbert uses sociological models to examine the link between community and collective action in south Wales and Nottinghamshire.[30] He argues that in the south Wales industrial villages the SWMF lodges effectively marginalised the supra-regional coal companies in the local community, while in Nottinghamshire the smaller owners retained their power base.[31] The development and structure of these local societies were of paramount importance for sustaining the collective action necessary to maintain the lengthy strikes of the 1920s. However, although he includes chapters on the south Wales and the Nottinghamshire coalfields in general, the work remains essentially a microstudy of Ynysybwl and Hucknall.

Roger Fagge and Werner Berg have both attempted explicitly cross-regional and cross-national comparatives. In his *Power, Culture and Conflict in the Coalfields*, Fagge investigates patterns of protest in the mining communities of West Virginia and south Wales between 1900 and 1922. The work has some parallels with Gilbert's in that it stresses the importance of the 'nature of culture and community' and the power relations within them as explanations for patterns of protest.[32] He sees the West Virginia communities as culturally fragmented along lines of ethnicity and religion, a feature encouraged by coalowners.[33] On the other hand the issues of ethnicity and religion were not so divisive in south Wales.[34]

Werner Berg, while noting the importance of social factors, concentrates upon the differing patterns of industrialisation as an explanation for the constellation of different interest groups, both of employers and workers.[35] Berg argues that the different industrial formations led to different proletarian formations. The industrial pit village in south Wales led to a 'purer type of working-class' as various aspects of workers' lives were under the control of the workers themselves. On the other hand, the more urbanised communities of the Ruhr miners were subject to 'foreign social (and state) control'.[36] This work shared similarities with Gilbert in that it stressed the control the miners had of the social arena in south Wales compared to the lack of similar influence in the Ruhr. However, Berg's work leaves less room for individual agency than Gilbert's does. The relatively greater influence of the workers themselves in the Welsh pit villages was the result of the distinctive pattern of industrialisation that shaped those communities. In this light Berg's work seems somewhat reductionist. However, it does recognise the influence that institutions that stood outside the labour movement could have upon workers' attempts at combination.

Both Fagge's and Berg's works remain anchored in structural explanations. While the importance of these structural explanations is apparent, it is the contention of this book that they do not provide a complete picture. The social structures of the mining communities in another American coalfield, Pennsylvania, were as diverse as (if not more so than) those of West Virginia and the Ruhr. Yet despite this, one dominant union in the form of the United Mineworkers of America (UMA) was able to emerge.[37] This suggests that theoretical frameworks that rely on social diversity alone as a means of explanation of trade union and political development provide an incomplete picture.

Finally, the book is specifically interested in the language used by the trade unions and political parties in the two coalfields. Welsh historiography has been somewhat sheltered from the debates that have accompanied the rise of the 'linguistic turn'. These debates have often revolved around concepts taken as given by earlier labour historians, such as class.[38] However, it should be pointed out that historians of south Wales have addressed the issue of political language, although they have not specifically engaged with the arguments of poststructuralists.[39] German labour history has also been somewhat sheltered from the 'linguistic turn'. For decades it has been dominated by a functionalist approach which consigns the German working class to passivity,[40] although *Alltagsgeschichte* has encouraged a more activist view of workers.

Although hotly contested, the work of both Gareth Stedman Jones and Patrick Joyce has served to heighten historians' appreciation of the role language and discourse played in creating identities like class.[41] Joyce, for example, has recognised the important role trade unions played in 'creating popular conceptions of the social order, particularly class perceptions. Union discourse was more than the reflection of the workers' worlds. It actively shaped it.'[42] However, although the present work is concerned with discourse, it does not go as far as Joyce in disassociating it from materialism.[43] More recent work has tried to find a medium between structure and discourse, while others have argued for an integration of the 'linguistic turn' and the older social history.[44] It is my contention that these approaches are not inimical, but can be fruitfully combined. As Jon Lawrence argues, historians must be aware of the 'basic parameters of political activity within communities', such as working practices, residential patterns and the composition of workforces. [45] Yet at the same time we need to be aware of the role organisations played in 'constructing political unity among fragmented groups of workers'.[46]

This work argues that the trade unions and political parties in the south Wales and Ruhr coalfields tried to construct that unity through discourse. I argue that these discourses were in turn constitutive of relations between organisations. This may seem to skirt perilously on the verge of linguistic determinism. But in my understanding, discourses, be they Marxist, anarchist or Christian-socialist, did not provide a rigid template for action. Rather they offered a framework for understanding events. They did not and could not dictate a single course of action, for all too often the 'parameters of political action' intervened, but they could suggest a preferred alternative. Structural and material factors set those parameters. They provided the potentiality for a discourse, but it took the trade unionists and the politicians to form it into a political language.

## Lifeworld, constructed identities and civil society

Miners must be the single group of workers subject to the most academic scrutiny. The 'occupational community', the 'isolated mass',[47] 'separatist group',[48] the 'traditional mining community',[49] the 'independent collier',[50] and a modified 'isolated mass'[51] have all been deployed as potential paradigms through which to view the miners and their collective action. It would be tedious to outline the merits and limitations of all these approaches; yet the question remains, how can we integrate these theories and the existing historiographies? This work attempts to bring these ideas together through three interrelated concepts: Jürgen Habermas's concept of the lifeworld, constructed identity and civil society.

Habermas argues that the lifeworld is made up of 'more or less, diverse background convictions'.[52] The lifeworld, therefore, represents a stock of interpretative patterns, a set of common assumptions that cannot be transcended.[53] Moreover this 'cultural tradition shared by a community is constitutive of the lifeworld, which the individual member finds already interpreted. This intersubjectively shared *lifeworld* forms the background for communicative action.'[54] Habermas sees the lifeworld as intersubjective since others inhabit it. The lifeworld provides the set of common assumptions, which provides the background for individuals when they engage in communicative action.

The links between milieux and the lifeworld might at first seem unclear. Habermas has insisted that the lifeworld cannot be used to define social stratification. However, Stefan Berger, in his comparison of south Wales and the Ruhr, has synthesised both concepts. He argues that the existence of various milieux in the Ruhr entailed a more fragmentary lifeworld. On the other hand, the more synchronous lifeworld of the Welsh miner encouraged a greater organisational unity.[55] However, Berger's work is based mainly on secondary literature. This book applies this interpretation to the primary source material for both areas. Examining how trade unions and political parties engaged in communicative action as organisations offers a fruitful avenue through which to examine the divergent developmental tendencies in south Wales and the Ruhr. This approach has the advantage of acknowledging the importance of the structural context in which organisations were formed, but also emphasises the role union and party activists played in their creation and maintenance.

Leading on from the lifeworld, the present work seeks to approach the communicative action of the trade unions through an examination of their discursive construction of identity. Christopher Johnson has examined the link between lifeworld and discourse. Writing from a materialist perspective, Johnson acknowledges criticisms of economic determinism. However, he also warns of the dangers of the 'linguistic turn'. He suggests that Habermas's concept of lifeworld provides a way of linking discourse, workers' culture and the structural forces that shape it.[56] As I argued above, discourse is presented here as an on-going effort to understand and interpret the world. Attempts at 'making sense' raise the possibility of myriad, sometimes conflicting, sometimes complementary, interpretations. Thus, organisational fragmentation or homogenisation was not simply a reflection of the social divisions or the lack of them. In fact, there was a more complex, dynamic relationship between societal structures and organisations.

The concept of identity constructed through discourse recognises the trade union activists as active in creating such an identity: an ideal of the miner to which they aspired. Such identities were used as a banner for recruitment and an ideological weapon. In some respects, this concept has similarities to Roy Church and Quentin Outram's 'constructed solidarity',[57] but while they were mainly concerned with explaining collective action during strikes, constructed identity has wider ramifications. These constructed identities often held up an image of the miner as he should be and can be seen, as stemming from the wider spheres of both work and community. They represent attempts by the trade unions and political parties to create unity among their potential supporters, by subsuming individual differences under a common, overarching identity. By utilising the idea of constructed identity the book also returns some agency to the activists of the political parties and the unions.

Habermas also provides the link with the final explanatory factor used in this book: the nature of civil society in Britain and Germany during this period. Geoff Eley argues that in modified form Habermas's idealised concept of the public sphere provides a useful analytical tool for the historian. Instead of the ideal public sphere, where rational communication could be pursued on a free and equal participatory basis, Eley suggests that it was a contested site. Access was denied or restricted on grounds of class, ethnicity, religion and gender. For Eley this competition involved issues of power.[58] As we shall see, the various trade unions and political parties examined in this work can, therefore, be seen as engaged in competition to dominate this public sphere.

However, the trade unions and political parties in the two coalfields were not acting in a vacuum. The way they constructed and expressed identity was influenced by the nature of the civil society in which they operated. Nancy J. Rosenblum and Robert C. Post define civil society as the realm of pluralism, but argue that 'we can in fact range pluralist societies along a spectrum from the fluid to the rigidly segmented'. The crucial feature of a fluid civil society is that the membership of more than one group is accepted. On the other hand, in a segmented civil society 'groups are more inclined to see membership as mutually exclusive and to be hostile to the idea of plural identities and multiple, overlapping membership'.[59] Significantly, Rosenblum and Post also argue that 'within civil society identity is always specific. Identity springs from contingencies of socialization and from choice; it is not the product of a single and unified design. It comes from participation in particular groups and associations, and is not imposed uniformly upon all by the state.'[60] This emphasis on the role that voluntary group membership plays in forming identity is important in that it avoids the danger of sociological determinism inherent in the concept of the milieu. This definition of civil society and identity therefore allows the historian to accommodate both structural factors and agency.

Concepts of civil society (and the public sphere) have already been applied to both British and German history. In the introduction to *Paradoxes of Civil Society*, Frank Trentmann, like Eley, focuses on issues of power, of exclusivity and inclusivity.[61] This book is not intended as an analysis of German and British civil societies, but takes as a starting point the suggestion in *Paradoxes of Civil Society* that Germany tended towards a segmented civil society, while in Britain society was more fluid. In Germany the boundary between civil society and the state were blurred. Daniel A. Macmillan has argued that parties and associations in Wilhelmine Germany acted under the assumption that the state and civil society should coexist in harmony and that civil society should be harmoniously united behind a single party. Each group's expectation of harmony made organisations intolerant of pluralist debate.[62] On the other hand, Trentmann's examination of the free trade movement in Britain focuses on its belief in the separation of civil society and the political system.[63]

This work aims to show how the discursive practices of the trade unions and political parties were crucial to reinforcing and perpetuating the processes of homogenisation in south Wales and fragmentation in the Ruhr. Using the threefold explanation outlined here (lifeworld, constructed identity and the nature of civil society) the work seeks to integrate both concerns of structure and discourse. The lifeworld provides the cultural background to the discursive construction of identity by trade unions and political parties, while those discourses were mediated through the different nature of civil societies. Two of the subsequent chapters, therefore, concentrate on the issue of lifeworld and civil society. The remaining three chapters examine the discourse of the trade unions and parties. They focus mainly on materials published by the organisations, such as newspapers and leaflets, and the speeches made by activists. Since identities are more strongly drawn during periods of conflict, I have focused on literature produced during industrial disputes and elections. Generally, I have concentrated on literature produced within the regions. Unfortunately, communist newspapers in the Ruhr were often banned in the 1920s, and those that do exist are fragmentary. Therefore, much of the examination of the Communist Party of Germany (KPD) discourse relies on flyers and handbills.

Finally, in the case of the Ruhr, the work focuses on discourse of the two main German trade unions and largely excludes the Polish union. This is entirely due to the failings of the author. An examination of the Polish union's discourse awaits a better linguist.

Critics will point out that this reliance on trade union literature and political speeches will inevitably mean that the work largely deals with the discourse of elites, since 'by definition the world-view of the activist is not the same as the non-activist'.[64] This is a justifiable criticism, but this book deals with organisational identities, not social identities, and it was the activists who were the crucial figures in shaping the identity of the associations discussed here. Also, given that the 'world-view' of the non-activist is so difficult to access, we cannot be entirely sure that it *never* corresponded to that of the activist. It seems more likely that those views had a dynamic relation, changing in light of circumstance, sometimes corresponding, sometimes diverging.

Chapter 1 provides a brief history of the socio-economic development of the two coalfields up to 1914 and sets the parameters in which the miners' movements operated. It contrasts the synchronous nature of the Welsh miners' lifeworld with the fragmented lifeworld of the Ruhr miners. It also highlights the greater fluidity of civil society in south Wales compared to the milieux of the Ruhr. In order to systemise this, the chapters applies M. I. Bulmer's eight-point model of the 'traditional mining community', which consists of: the degree of physical isolation of communities; the economic predominance of mining; whether the job was a source of pride; the degree of occupational homogeneity and isolation; the extent of communal activities with workmates; the sharpness of gender divisions and generational continuity in occupation; the extent of conflict with owners; a combination of all these factors.[65]

Chapter 2 examines the development of the various trade unions before 1914. Emphasis here is placed on how the SWMF created a comparatively inclusive, flexible identity through its toleration of internal debate. In the Ruhr, on the other hand, the attempts to enforce harmony within the trade unions resulted in the creation of exclusive, more doctrinal identities.

Chapter 3 turns to the political sphere and focuses on the discursive construction of identities by political parties. It emphasises the changing meaning attached to the notion of class within south Wales, while indicating how the parties in the Ruhr provided contested notions of the German *Volk*.

Chapter 4 examines how the Great War impacted on the lifeworld in the two regions, and thereby altered the parameters in which the organisations operated. It argues that in south Wales there was a greater continuity in the lifeworld, while in the Ruhr there was increased fragmentation through mechanisation and rationalisation. In general it also shows that the Welsh miners made some significant gains during the war, while the Ruhr miners were unable to capitalise on labour shortages.

Chapter 5 combines an examination of both trade union and party political discourse through an examination of the key issues of nationalisation and socialisation. While nationalisation received broad support among the Welsh miners, in the Ruhr the idea of socialisation was contested by different sections of the labour movement.

Finally, the conclusion brings together the main arguments to provide a revised understanding of the nature of trade union and party political development in the two regions, one which accepts the important role of both structural factors and activists' discourse.

## Notes

1 MFGB, Messrs Robert Smillie and Alfred Onions Report on the Social and Industrial Conditions of the German Working Class in 1910, TUC HD6661.M6.19.

2 G. Werner, *Was kann der deutsche Bergbau von England lernen?* (Essen: Fredebeul & Koenen, 1910).

3 For Britain see R. P. Arnot, *The Miners: A History of the Miners' Federation of Great Britain*, 3 volumes (London: Allen & Unwin, 1949) and Arnot, *Miners: One Union, One Industry: A History of the National Union of Mineworkers* (London: Allen & Unwin, 1979). For Germany see O. Hue, *Die Bergarbeiter. Historische Darstellung der Bergarbeiter-Verhältnisse von der Ältesten bis in die neueste Zeit* (Berlin: J. H. W. Dietz, 1913/1981) and H. Imbusch, *Arbeitsverhältnis und Arbeiter-Organisationen im deutschen Bergbau: Eine geschichtliche Darstellung* (Berlin/Bonn: J. H. W. Dietz, 1908/1980).

4 A. Horner and N. Watkins, *British Mineworkers' Union: A Constitution*, 1926, TUC HD6350, pp. 1–2.

5 See for example, J. Tampke, *The Ruhr and Revolution: The Revolutionary Movement in the Rhenish-Westphalian Industrial Region 1912–1919* (London: Croom Helm, 1979) and W. Paynter, 'The "Fed"', in Goronwy Alun Hughes (ed.), *Men of No Property: Studies of Welsh Trade Unionism* (Caerwys: Gwasg Gwenffrwd, 1971), pp. 68–72.

6 D. Geary, 'The myth of the radical miner', in S. Berger, A. Croll and N. LaPorte (eds), *Towards a Comparative History of Coalfield Societies* (Aldershot: Ashgate, 2005), pp. 43–64.

7 For an overview of these traditions see S. Berger and N. Evans, 'Two faces of King Coal: the impact of historiographical traditions on comparative history in the Ruhr and south Wales', in Berger, Croll and LaPorte (eds), *History*, pp. 29–42.

8 H. Francis and D. Smith, *The Fed: A History of the South Wales Miners in the Twentieth Century*, 2nd edn (Cardiff: University of Wales Press, 1998), p. xvi.

9 C. Williams, *Democratic Rhondda: Politics and Society, 1885–1951* (Cardiff: University of Wales Press, 1996), pp. 7–9; M. Lieven, *Senghennydd: The Universal Pit Village, 1890–1930* (Llandysul: Gomer Press, 1994).

10 See A. Croll, *Civilizing the Urban: Popular Culture, Public Space and Urban Meaning in Merthyr, c. 1870–1914* (Cardiff: University of Wales Press, 2000); A. V. John (ed.), *Our Mothers' Land: Chapters in Welsh Women's History, 1830–1939* (Cardiff: University of Wales Press, 1991); A. V. John, 'A miner struggle? Women's protests in Welsh mining history', *Llafur*, 4:1 (1984), 72–90; D. Jones, 'Serfdom and slavery: women's work in Wales 1890–1930', in D. R. Hopkin and G. S. Kealey (eds), *Class, Community and the Labour Movement: Wales and Canada* (Aberystwyth: Llafur/CCLH, 1989), pp. 86–100 and C. White and S. R. Williams (eds), *Struggle or Starve: Women's Lives in the South Wales Valleys Between the Two World Wars*

(Dinas Powys: Honno, 1998). In fact Dai Smith has also been a prominent figure in the study of popular culture in south Wales. See D. Smith, *Wales! Wales?* (London: Allen & Unwin, 1984); D. Smith, *Aneurin Bevan and the World of South Wales* (Cardiff: University of Wales Press, 1993).

11 See for example Tampke, *Ruhr and Revolution*, pp. 160–1. See also J. Tampke, 'Die Sozialisierungbewegung im Steinkohlenbergbau an der Ruhr', in H. Mommsen and U. Borsdorf (eds), *Glück auf, Kameraden. Die Bergarbeiter und ihre Organisationen in Deutschland* (Cologne: Bund-Verlag, 1979), pp. 225–48.

12 K. Hartewig, *Das unberechenbare Jahrzehnt. Bergarbeiter und ihre Familien im Ruhrgebiet 1914–1924* (Munich: Beck, 1993), pp. 318–21.

13 K. Tenfelde, *Sozialgeschichte der Bergarbeiterschaft an der Ruhr im 19. Jahrhundert* (Bonn: Verlag Neue Gesellschaft, 1977). See also K. Tenfelde, 'Der bergmännische Arbeitsplatz während der Hochindustrialisierung 1890–1914', in W. Conze and U. Engelhardt (eds), *Arbeiter im Industrialisierungsprozeß* (Stuttgart: Ernst Klett, 1979), pp. 283–335.

14 S. Hickey, *Workers in Imperial Germany: The Miners of the Ruhr* (Oxford: Clarendon Press, 1985).

15 D. F. Crew, *Town in the Ruhr: A Social History of Bochum, 1860–1914* (New York: Columbia University Press, 1979), pp. 186–93. The notion of occupational community is taken from G. Salaman, 'Some sociological determinants of occupational communities', *Sociological Review*, 19:1 (1971), 53–77.

16 D. F. Crew, '*Alltagsgeschichte*: A new social history "from below"?', *CEH*, 22:3/4 (1989), 394–407. For more on the debate between *Strukturgeschichte* and *Alltagsgeschichte* see R. Fletcher, 'History from below comes to Germany: The new history movement in the Federal Republic of Germany', *JMH*, 60:3 (1988), 557–68; W. L. Patch, 'German social history and labor history: A troubled partnership', *JMH*, 56:3 (1984), 483–98; P. Schöttler, 'Historians and discourse analysis', *History Workshop* 27, spring (1989), 37–68; G. Eley, 'Labor history, social history, *Alltagsgeschichte*: Experience, culture, and the politics of the everyday – a new direction for German social history', *JMH*, 61:2 (1989), 296–343.

17 F. Brüggemeier, *Leben vor Ort. Ruhrbergleute und Ruhrbergbau, 1889–1919* (Munich: Beck, 1984), p. 79.

18 *Ibid.*, pp. 136–41. See also D. Peukert, 'Industrialisierung des Bewußteins? Arbeitserfahrugen von Ruhrbergleuten im 20. Jahrhundert', in K. Tenfelde (ed.) *Arbeit und Arbeitserfahrung in der Geschichte* (Göttingen: Vandenhoeck and Ruprecht, 1986).

19 See F. Brüggemeier and L. Niethammer, 'Schlafgänger, Schnapskasinos und schwerindustrielle Kolonie. Aspekte der Arbeiterwohnungsfrage im Ruhrgebiet vor dem Ersten Weltkrieg', in J. Reulecke and W. Weber (eds), *Fabrik, Familie, Feierabend. Beiträge zur Sozialgeschichte des Alltags im Industriezeitalter* (Wuppertal: Hammer, 1978), pp. 135–76; F. Brüggemeier, 'Soziale Vagabundage oder revolutionär Heros? Zur Sozialgeschichte der Ruhrbergarbeiter 1880–1920', in L. Niethammer (ed.), *Lebenserfahrung und kollektive Gedächtnis. Die Praxis der >>Oral History<<* (Franfurt am Main: Syndikat, 1980), pp. 193–213.

20 Brüggemeier, *Leben vor Ort.* For a concise overview of these various accounts of miners' unionism in the Ruhr see G. Eley, 'Labor history, social history, *Alltagsgeschichte*', pp. 328–35.

21 Eley, 'Labor History, social history, *Alltagsgeschichte*', pp. 296–343.

22 See for example, K. O. Morgan, *Wales in British Politics, 1868–1922* (Cardiff:

University of Wales Press, 1980).

23 Readers may compare Morgan, *Wales in British Politics* to Williams, *Democratic Rhondda* as indicative of the move to the microhistorical. There are many articles examining the rise of Labour within individual constituencies. See for example, D. Hopkin, 'The rise of Labour: Llanelli, 1890–1922', in G. H. Jenkins and J. B. Smith (eds), *Politics and Society in Wales, 1840–1922* (Cardiff: University of Wales Press, 1988), pp. 161–82.

24 See the various essays in D. Tanner, C. Williams and D. Hopkin, *The Labour Party in Wales, 1900–2000* (Cardiff: University of Wales Press, 2000).

25 W. Claggett, J. Loesch, W. P. Shively and R. Snell, 'Political leadership and the development of political cleavages: Imperial Germany, 1871–1912', *AJPS*, 26 (1982), 643–64.

26 R. Lepsius, 'Parteiensystem und Sozialstruktur. Zum Problem der Demokratisierung der Deutschen Gesellschaft', in G. A. Ritter (ed.), *Die Deutschen Parteien vor 1918* (Cologne: Kiepenhauer & Witsch, 1973), pp. 56–80.

27 W. Jäger, *Bergarbeitermilieus und Parteien im Ruhrgebiet. Zum Wahlverhalten des katholischen Bergarbeitermilieus bis 1933* (Munich: Beck, 1996).

28 See K. Rohe, W. Jäger, U. Dorow, 'Politische Gesellschaft und politische Kultur' in Wolfgang Köllmann *et al.* (eds), *Das Ruhrgebiet im Industriezeitalter. Geschichte und Entwicklung,* vol. 2 (Düsseldorf: Schwann im Palmos-Verlag; 1990), K. Rohe, 'Katholiken, Protestanten und Sozialdemokraten im Ruhrgebiet vor 1914. Voraussetzungen und Grundlagen "konfessionellen" und "klassenbewußten" Wählens in einer Industrieregion', in K. Rohe, *Vom Revier zum Ruhrgebiet. Wahlen, Parteien, Politische Kultur* (Essen: Hobbing, 1986), pp. 43–59.

29 See P. Cooke, 'Radical regions? Space, time and gender relations in Emilia, Provence and South Wales', in G. Rees *et al.* (eds), *Political Action and Social Identity: Class, Locality and Ideology* (London: Macmillan, 1985), pp. 17–41 and P. Cooke, 'Class practices as regional markers: A contribution to Labour geography', in D. Gregory and J. Urry (eds), *Social Relations and Spatial Structures* (London: Macmillan, 1985), pp. 212–41.

30 D. Gilbert, *Class, Community and Collective Action: Social Change in Two British Coalfields, 1850–1926* (Oxford: Clarendon Press, 1992). The theories considered are Kerr and Siegal's isolated mass theory, Mancur Olson's theory of small groups and Craig Calhoun's communal collective action. He provides a full discussion of these approaches in chapter 2.

31 *Ibid.*, pp. 247–51.

32 R. Fagge, *Power, Culture and Conflict in the Coalfields: West Virginia and South Wales, 1900–1922* (Manchester: Manchester University Press, 1996), p. 6.

33 *Ibid.*, pp. 66–72.

34 *Ibid.*, pp. 77–83.

35 See W. Berg, *Wirtschaft und Gesellschaft in Deutschland und Großbritannien im Übergang zum 'organisierten Kapitalismus'. Unternehmer, Angestellte, Arbeiter und Staat im Steinkohlenbergbau des Ruhrgebietes und von Südwales, 1850–1914* (Berlin: Duncker und Humblodt, 1984), and W. Berg, 'Zwei Typen industriegesellschaftlicher Modernisierung: Die Bergarbeiter im Ruhrgebiet und in Südwales im 19. Jahrhundert und frühen 20. Jahrhundert', in Gustav Schmidt (ed.), *Bergbau in Großbritannien und im Ruhrgebiet. Studies zur vergleichen Geschichte des Bergbaus 1850–1930* (Bochum: Bockmeyer, 1985).

36 Berg, 'Zwei Typen', p. 218.

37 P. Blatz, *Democratic Miners: Work and Labour Relations in the Anthracite Coal Industry 1875–1925* (Albany: University of New York Press, 1994); H. Aurand, *From Molly Maguires to the United Mineworkers* (Philadelphia: Temple University Press, 1971).

38 A. Croll, '"People's Remembrances" in a post-modern age: contemplating the non-crisis of Welsh Labour History', *Llafur*, 8:1 (2000), 5–17.

39 See P. Stead, 'The language of Edwardian politics', in David Smith (ed.), *A People and a Proletariat: Essays in the History of Wales, 1780–1980* (London: Pluto Press, 1980), pp. 148–65; Williams, *Democratic Rhondda*, pp. 7–9.

40 R. J. Evans, 'Introduction: the Sociological Interpretation of German Labour History', in R. J. Evans (ed.), *The German Working Class 1883–1933: The Politics of Everyday Life* (London: Croom Helm, 1982), pp. 15–45.

41 G. Stedman Jones, *Languages of Class: Studies in English Working Class History, 1832–1982* (Cambridge: Cambridge University Press, 1983); P. Joyce, *Visions of the People: Industrial England and the Question of Class, 1848–1914* (Cambridge: Cambridge University Press, 1991).

42 Joyce, *Visions of the People*, p. 137.

43 P. Joyce, 'Refabricating labour history; or, from labour history to the history of labour', *LHR*, 62:2 (1997), 147–52.

44 N. Kirk and J. Belchem (eds), 'Introduction', *Languages of Labour* (Aldershot: Ashgate, 1997); J. Lawrence, *Speaking for the People: Party, Language and Popular Politics in England 1867–1914* (Cambridge: Cambridge University Press, 1998); J. Vernon, *Politics and the People: A Study in English Political Culture, 1815–1867* (Cambridge: Cambridge University Press, 1993). For an overview see A. Croll, 'The impact of postmodernism on British social history', *Mitteilungsblatt*, 27 (2002), 137–52; D. Reid 'Reflections on labor history and language', in L. R. Berlanstein (ed.), *Rethinking Labor History: Essays on Discourse and Class Analysis* (Chicago: University of Illinois Press, 1993), pp. 39–54.

45 Lawrence, *Speaking for the People*, p. 64.

46 *Ibid.*, p. 66.

47 C. Kerr and A. Siegel, 'The interdisciplinary propensity to strike – an international comparison', in R. Dubin, A. Kornhauser and A. M. Ross (eds), *Patterns of Industrial Conflict* (New York, Toronto, London: McGraw Hill, 1954), pp. 189–212.

48 G. V. Rimlinger, 'International differences in the strike-propensity of coal miners: Experience in four countries', *Industrial and Labour Revolutions Review* (1959), 389–405.

49 M. I. Bulmer, 'Sociological models of the mining community', *Sociological Review*, 23:1 (1975), 83. Bulmer also provides a general overview of the isolated mass and separatist group theories.

50 R. Harrison, *Independent Collier: The Coal Miner as Archetypal Proletarian Reconsidered* (Sussex: Harvester Press, 1978)

51 R. Church and Q. Outram, *Strikes and Solidarity: Coalfield Conflict in Britain, 1889–1966* (Cambridge: Cambridge University Press, 1998), pp. 132–5.

52 J. Habermas, *The Theory of Communicative Action: Reason and the Rationalization of Society*, translated by T. McCarthy (Cambridge: Polity Press, 1984), p. 70.

53 J. Habermas, *The Theory of Communicative Action: The Critique of Functionalist Reason*, translated by T. McCarthy (Cambridge: Polity Press, 1987), pp. 131–3.

54 Habermas, *Theory of Communicative Action: Reason and the Rationalization*, p. 82.

55 S. Berger, 'Working-class culture and the labour movement in the south Wales and

the Ruhr coalfields: A comparison', *Llafur*, 8:2 (2001), 5–40.

56 C. H. Johnson, 'Lifeworld, system, and communicative action: The Habermasian alternative in social history', in Berlanstein (ed.), *Rethinking Labor History*, pp. 55–89.

57 Church and Outram, *Strikes and Solidarity*, p. 115.

58 G. Eley, 'Nations, publics, and political cultures: Placing Habermas in the nineteenth century', in Craig Calhoun (ed.), *Habermas and the Public Sphere* (Cambridge, Mass. and London: MIT Press, 1992), pp. 289–339.

59 N. J. Rosenblum and R. C. Post (eds), 'Introduction', in N. J. Rosenblum and R. C. Post (eds), *Civil Society and Government* (Princeton and Oxford: Princeton University Press, 2002), pp. 4–5.

60 *Ibid.*, p. 7.

61 F. Trentmann, 'Introduction', in F. Trentmann (ed.), *Paradoxes of Civil Society. New Perspectives on Modern German and British History* (New York and Oxford: Berghahn Books, 2000), pp. 3–46.

62 D. A. Macmillan, 'Energy, willpower and harmony: On the problematic relationship between state and civil society in nineteenth-century Germany', in F. Trentmann (ed.), *Paradoxes of Civil Society*, pp. 176–98.

63 F. Trentmann, 'Civil society, commerce, and the "citizen-consumer". Popular meanings of free trade in modern Britain', in F. Trentmann (ed.), *Paradoxes of Civil Society*, pp. 306–31.

64 Lawrence, *Speaking for the People*, p. 66.

65 Bulmer, 'Sociological models', pp. 85–9.

# 1

# The growth of the coal industries, 1850–1914

> ... you could look at every door, and you'd see a miner coming out. And that miner and his family would know every person that lived in the village. They went to the same chapels, or to the same pub, and they always met in the pub and in the chapel.
>
> Thomas Nicholas, ex-miner, talking about the south Wales mining communities.[1]

> Everywhere one sees Slavic faces and hears [Slavic] voices beat against one's ear.
>
> Lorenz Pieper in his examination of the mining communities in the Ruhr.[2]

## Introduction

This chapter examines the development of the coal industries in both the Ruhr and south Wales up to 1914. It aims to investigate the social structures that provided the backdrop against which the trade unions and political parties worked. By examining the social context in which activists operated it also seeks to throw light on the nature of the lifeworld in the two areas. First, it examines the growth of the industries through the nineteenth century and into the twentieth, where the necessary capital investment came from, the markets that production supplied and the organisation of the employers. Secondly, it investigates the miners' experiences of work and the potential that existed for conflict and for expressions of solidarity. Finally, the social structures of the mining communities and the types of civil societies are examined and compared with reference to the degree of cohesiveness they exhibited.

## King coal: the expansion of mining

Among this frenetic activity the opening of the coalfields of south Wales and the Ruhr stand out as paragons of industrialisation. The latter formed part of Germany's generally swift industrialisation in the latter half of the nineteenth century. Meanwhile, although Britain as a whole experienced a far longer period of industrialisation, the growth of the south Wales coalfield was remarkable in that, from its beginnings in the 1850s, it was to achieve a dominant position in the world coal trade. Its exports alone constituted almost 33 per cent of total world coal exports between 1890 and 1910.[3]

Small-scale coal mining had been pursued in south Wales and the Ruhr prior to the massive investment of capital that the second half of the nineteenth century introduced. In south Wales, coal production in the first half of the nineteenth century largely fuelled the growth of the metal industries centred on Merthyr.[4] Similarly, early Ruhr coal production fuelled the small iron industries along the Ruhr and the growing industrial empire of the Krupp dynasty in Essen.[5]

Growth was prompted by technological advances in ventilation and drainage which allowed access to deeper seams through deep mines. In south Wales mining moved from the slopes of the valleys to the valley floors, while in the Ruhr capitalists were able to begin exploitation of the seams in the Emscher basin, north of the river Ruhr.[6] The industry in the Ruhr can be divided into several zones. The oldest ran along the banks of the river Ruhr itself. From the 1840s mining works advanced northwards to embrace the Hellweg, Emscher and Vest.[7] In south Wales, the sale-coal industry initially developed around Aberdare from 1837 to 1862 before expanding into the Rhondda and the Garw in the 1870s. By 1900 the Sirhowy, Bargoed, Rhymney and Maesteg areas were the frontiers of the industry.[8] By 1913 the number of mines had reached 485 with a workforce of 233,134. In the Ruhr there were 167 mines with a workforce of 394,569 in the same year.[9] Comparison of the regions' total output of coal demonstrates the expansion of the industry (see Table 1.1).

*Table 1.1* Average output of coal in the Ruhr and south Wales per annum

| Ruhr | | South Wales | |
|---|---|---|---|
| Year | Output (in 000 tons) | Year | Output (millions of tons) |
| 1888–1892 | 35,511 | 1891–1895 | 31.5 |
| 1908–1912 | 88,317 | 1911–1914 | 52.8 |

*Source:* N. J. G. Pounds, *A Historical Geography of Europe, 1800–1914* (Cambridge: Cambridge University Press, 1985), p. 362; Walters, *Economic and Business History of South Wales Coal Industry* (New York: Arno Press, 1977), p. 6.

In both areas the expansion of the industry was in private hands. In Britain the notions of *laissez-faire* economic activity were firmly established by the time the south Wales coalfield was opened. On the other hand, in the Ruhr government control over the coalmines was only relinquished in 1865. Previously, the mines had been run under the principle of the *Direktionsprinzip*, whereby the state officials controlled conditions and wages. In neither area, however, were capitalists left with an entirely free hand. The Prussian State Mining Office (*Oberbergamt*), the body responsible for ensuring that the mining regulations were observed by the private companies, survived the liberalisation of the 1860s, while the various Acts of Parliament illustrate that successive British governments were willing to intervene in the industry.[10]

The capital investment necessary for new pits came from a variety of sources. In south Wales the most common source were consortia or partnerships of wealthy individuals. In the Ruhr, capital was largely raised during the first half of the nineteenth century from small, local investors, but following the abolition of the *Direktionsprinzip* and the development of more expensive technologies banks increasingly became a main source of capital. In both areas increasing costs led joint-stock companies (*Aktiengesellschaft*) gradually to displace family concerns although some family concerns, such as Krupp, retained large coal interests.[11] The establishment of these companies has been regarded as vitally important for the emergence of Germany as a major European economic force during the nineteenth century, as they facilitated the modernisation of certain sectors, including coal and iron production.[12] However, the sale of shares to create these joint-stock companies should not be regarded as creating some form of share-holding 'democracy'. 'Public subscription', that is the purchasing of shares by the public at large, was rare, as shares were sold and bought in large denominations beyond the reach of small investors.[13] Both Welsh and German industrialists also generated investment by ploughing back profits into their enterprises.[14]

The coal produced by these two industries generally found its way to different markets. The expansion of the industry in south Wales marked a departure from previous production as coal became an export commodity in itself. John Nixon, who later became a major coalowner in his own right, first arrived in south Wales in 1839 to seek an appointment at Crawshay Bailey's combined ironworks and colliery.[15] On his second visit Nixon came specifically seeking coal, 'a supply of black diamonds' for which he was convinced he had found a market in France.[16] Production shifted from serving the local ironworks to enterprises based on sale-coal, although some firms, such Guest, Keen and Nettleford of Dowlais, continued to operate mines to supply their metalworks. As an exporting region the south Wales industry benefited from its proximity to the Bristol Channel and favourable freight rates. While Welsh coal needed only to be transported 25 miles to reach the point of export, Ruhr coal had to travel some 150 miles.[17] The Ruhr coal companies had access to river transport, but shipments of coal on the Ruhr actually decreased from the 1860s onwards.[18] Furthermore, although both geological areas contained different types of coal, much of south Wales's production was 'steam' coal and therefore capitalised upon the increasing use of steam power by shipping.

This export trade was facilitated by the growth of Newport, Cardiff and Barry docks.[19] The construction of these docks provides an example of how the coal industry in Wales ushered in and supported a host of ancillary industries such as shipping and railways. By 1913, 53 per cent of coal produced in south Wales was being exported.[20]

By contrast the Ruhr coal industry was far less export-oriented. Overseas export amounted to only about half a million tons by 1913, although there was trade with neighbouring France and Belgium.[21] Instead production largely served other indigenous heavy industry. More than 85 per cent of production was destined for internal markets.[22] The growth of the iron and steel industries in the Ruhr paralleled that of the coal industry. Initially, companies constructed forges next to mines that produced both coal and iron ore. Later ore was imported from Sweden. By 1914, iron and steel production amounted to 8.2 and ten million tons respectively.[23] The mines owned by metal magnates were referred to as *Hüttenzechen* and the close relationship between the two industries in the Ruhr can be illustrated through the example of the Krupp firm. It bought several small mines in the eighteenth century. After the company passed into the hands of energetic entrepreneur Alfred Krupp it expanded rapidly and extended its coal interests in order to fuel its production of railroad parts, machinery, shipping and armaments.[24]

This tendency towards concentration and integration of the industry was apparent in both regions throughout the nineteenth and twentieth centuries. By 1918 the Cambrian Combine, for example, controlled nine other colliery companies, apart from its own.[25] Meanwhile the Powell Duffryn and Ocean Coal Companies rose to become two of the largest producers through new sinkings and the purchase of smaller companies.[26] Vertical and horizontal concentration of industry was even more pronounced in the Ruhr. In 1885 only seven companies owned more than one mine; by 1910 nine were producing almost two-thirds of the region's total output. In both areas the number of small owners was shrinking in number, but with greater rapidity in the Ruhr (see Table 1.2). Table 1.3 reveals the much larger scale of mining operations in the Ruhr compared with south Wales. Three of the largest, the *Gelsenkirchener Bergwerks AG*, the *Harpener Bergbau AG* and *Bergwerksgesellschaft Hibernia*, had an output of 5 million tons each.[27]

Furthermore, in the Ruhr vertical integration of industry was encouraged by the growth of the joint-stock companies and links to the banks. Firms like Krupp or the *Dortmunder Union* who could 'build railways both at home and abroad entirely from its own resources, from the mining of the coal to the manufacture of track and the financing of the entire project' were typical.[28]

Cartelisation and syndication was also more pronounced in the Ruhr than in south Wales. The German employers entered into associations more readily than their Welsh counterparts. Even before the liberalisation of the economy there existed associations of employers in the west of the region such as the *Magerkohlenverband* in 1854.[29] Employers combined in 1858 to form the *Verein für die bergbaulichen Interessen im Oberbergamtsbezirk Dortmund* (Association for Mining Interests in Dortmund Mining District). Generally known as the *Bergbau Verein*, it acted as the spokes-body for the industry and was recognised as such by the government in 1893. The *Bergbau Verein* was followed by the formation of the *Rheinisch-Westfälische Kohlensyndikat* (RWKS) or Coal Syndicate in 1893. Members agreed to sell their coal to the RWKS who then handled the marketing of the product. It also imposed coal production quotas upon its members in an attempt to maintain price levels and moderate fluctua-

*Table 1.2* Output and number of employers in south Wales and the Ruhr coalfields, 1886–1916

| | Number of employers | | | |
|---|---|---|---|---|
| Output per year | South Wales | | Ruhr | |
| in thousand tons | 1886 | 1916 | 1893 | 1911 |
| 1–10 | 9 | 1 | | |
| 10–100 | 27 | 24 | 12 | |
| 100–500 | 24 | 45 | 71 | 22 |
| 500–1 million | 4 | 18 | 9 | 19 |
| Over 1 million | 3 | 14 | 6 | 26 |
| Total number | 68 | 101 | 98 | 67 |
| Average output | 202,481 t | 436,103 t | 368,457 t | 1,153,044 t |

*Table 1.3* Output in tons and size of the workforce of the four largest colliery companies in the Ruhr and south Wales in 1913

| | Output | No. of workers employed |
|---|---|---|
| *Ruhr* | | |
| Gelsenkirchener Bergwerks-Akteingesellschaft | 10,353,050 | 36,821 |
| Harpener Bergbau-Aktiengesellschaft | 8,635,577 | 30,987 |
| Hanielische Zechen | 6,677,621 | 21,539 |
| Bergwerksgesellschaft Hibernia | 5,697,596 | 19,314 |
| *South Wales* | | |
| Powell Duffryn Steam Coal Ltd | 3,873,780 | 13,611 |
| Consolidated Cambrian Ltd | 3,000,000 | 12,035 |
| Ocean Coal and Wilsons Ltd | 2,500,000 | 9,496 |
| Guest, Keen and Nettlefords Ltd | 2,000,000 | 10,487 |

*Source:* Berg, *Wirtschaft und Gesellschaft*, pp. 81 and 135.

tions in the market. These constraints encouraged larger firms to buy up smaller enterprises to increase their own production quotas, thereby contributing towards concentration. By 1904 a total of twelve mines had bought up for this reason, mostly older, southern portion of the Ruhr.[30] The power of the RWKS is revealed by the fact that between 1893 and 1913 its members were responsible for almost 90 per cent of total output between them.[31] However, its operation was not without problems as the *Hüttenzechen* were not covered and could sell their surplus upon the market, uncontrolled by the RWKS. It was only in 1903 that agreement to limit the output of coal to the RWKS quotas was reached.[32] The final employers' organisation, the *Zechen-Verband*, was founded in 1908 following the 1905 miners' strike. It was aimed at gaining greater control over the labour market through blacklists of troublesome employees, employment certificates and company controlled labour exchanges.[33]

In comparison to this formidable array of organisations, the employers of south Wales were rather weak. South Wales did have the third longest surviving owners' association in the form of the Monmouthshire and South Wales Coalowners' Association (MSWCOA), which had begun life in 1864 as the Aberdare Steam Collieries Association. Initially its members were responsible for 70 per cent of the region's coal production, but its membership was subject to wide fluctuations. Membership fell by 45 per cent when coal prices fell after 1876, only to rise again prior to the 1893 Hauliers' Strike.[34] Furthermore there was no equivalent in south Wales to the RWKS, despite the schemes to limit production being put forward by several individual coalowners. John Nixon proposed several quota systems between 1860 and 1885, only to be rebuffed by other employers.[35] In 1893 Sir George Elliot, in a plan that foreshadowed the creation of the National Coal Board, proposed the formation of a coal trust headed by nine men.[36] Later D. A. Thomas, head of the Cambrian Combine, advocated the allocation of quotas to colliery companies. However, neither scheme came to fruition. The disinterest of the majority of the south Wales owners in cartelisation was based more on the belief that it offered little advantage to the Welsh coalowners than an inherent attachment to *laissez-faire* economics. The price of their exports was determined by the price of coal from other coalfields despite internal competition. The MSWCOA, therefore, concerned itself mainly with labour relations, which largely meant resistance to any rises in overheads.[37] In this respect the MSWCOA can be regarded as broadly fulfilling the same functions as the *Zechen-Verband*.

## Management and work

In both areas early management was mainly the work of one or more major shareholders. The growth in size of operations led owners to become increasingly involved with the general management, while a stratum of salaried professional managers dealt with production at the pithead.[38] In this respect developments in south Wales and the Ruhr paralleled each other, although in the latter region a high proportion of mine managers came from the civil service.[39] Despite the economic liberalisation laws of the 1860s there remained close personal links between industry and government. Aristocrats, bureaucrats and politicians could, and did, have active roles within the industrial sphere.[40] The authoritarian stance of management was often coupled with a paternalistic ethos. For example, Otto Faust, an official at Neumühl, recalled that the company built its own club for managers since all the local pubs catered for miners.[41]

There were also comparatively more supervisory personnel in the Ruhr mines. In 1891 Her Majesty's Inspector of Mines for the South Wales District reported that in 1890 there were 205 managers and 177 undermanagers for 299 out of 354 mines.[42] In 1893 the number of managers (*Betriebsführer*) and undermanagers (*Obersteiger und Fahrsteiger*) in the Ruhr numbered 214 and 94 respectively.[43] However, by 1913 the Ruhr coal industry was employing an average of 233 officials per employer, an average of one technical official (*Angestellte*) per 28 workers. In south Wales the average was roughly one per 40 in 1914.[44] Furthermore, mines in the Ruhr tended to be deeper and have a larger workforce. Between 1875 and 1912/14 the average depth of mines in south Wales and the Ruhr increased to roughly 350 and 577 metres respectively. In 1913 the average number of men working at one pit was 383 and 1,815 respectively, while the size of total workforce per employer averaged 560 for south Wales and 4,365 for the Ruhr.[45]

Although technological advances allowed exploitation of deeper seams, corresponding advances in 'getting' coal were much slower. The mining industry of both areas remained very labour intensive and mechanisation was slow to have any significant impact. By 1908 there were 27 coal-cutting machines in use in collieries in the Cardiff and Newport District and 24 in the Swansea District. However, only four were recorded as having produced an appreciable output in the Swansea District.[46] A meagre one per cent of coal production in south Wales was machine cut in 1913. In comparison there were 237 machines (*Schrämmaschinen*) in 49 mines in the Ruhr by 1905.[47] The power hand drill had been introduced in 1906, but by 1914 there were fewer than 600 in use and, although the number of coal-cutting machines had risen to 449 by the previous year, they accounted for only 2.2 per cent of total production.[48] It has been suggested this low level of mechanised production in the Ruhr was due to uncoordinated investment and that the employers' reaction to adverse economic conditions was cartelisation rather than mechanisation.[49] Meanwhile, in south Wales, geological conditions militated against the use of machinery.[50] Bert Coombes, who moved from Herefordshire to south Wales in search of work, first encountered coal-cutting machinery during the First World War and wrote that they were 'rare in South Wales then, and there were none working in our valley or anywhere quite near'.[51] Therefore, while production in the Ruhr was clearly more mechanised than in south Wales, in neither area was mechanisation a major feature of coal winning prior to the Great War.

This should not suggest that employers and managers in both areas were insensible to the benefits of machinery. Articles advocating the benefits of machinery often appeared in the industry journals.[52] However, in both areas the expansion of the workforce took priority over the introduction of machinery. The production of coal, therefore, relied heavily upon the physical capabilities of the workers and horses. Of the latter there were 8,008 in the Ruhr and 16,744 in south Wales in 1913.[53] To compete companies had to expand their individual workforces. Table 1.4 indicates the increase in the mining labour force throughout the nineteenth century and into the twentieth.

Underground work generally followed the same lines in both regions. The miner would rise in the early morning and proceed to the mine with his packed lunch. There he would receive his lamp, wait for his turn to enter the cage and descend the shaft. From the shaft bottom he would proceed to his working-place (*Arbeitsstelle*) once the firemen (*Wettermänner*) had checked that the area was free from gas. In each working place the miners would work in small groups (*Kameradschaft*) of between three and five individuals. Generally, the miner would begin by undercutting the coal, an activity that was overwhelmingly

*Table 1.4* Estimates of number of miners employed in the south Wales and Ruhr coalfields, 1870–1913

| South Wales | | Ruhr | |
|---|---|---|---|
| Year | Number of miners | Year | Number of miners |
| 1871 | 57,000 | 1870 | 51,391 |
| 1891 | 115,600 | 1890 | 127,794 |
| 1913 | 232,400 | 1913 | 394,569 |

*Source:* Church, *History of the British Coal Industry*, p. 189, and Hickey, *Workers in Imperial Germany*, p. 13.

performed by hand with a pick or a mandrill. Where necessary the overhanging coal would be supported by wooden sprags. When ready they would be withdrawn allowing the undercut coal to fall. In some areas the coal was particularly prone to sticking to the top and had to be freed by the use of explosives. The coal would then be loaded into the trams, ready to be transported by the hauliers. For this work the hewers were paid on a piecework (*Gedinge*) system dependant on the amount of coal they produced.[54]

This brief description of the hewer's work obscures several differences. The hewer of the south Wales coalfield was more involved in deadwork, that is throwing the stone debris into the 'gob' and setting the supports. In the Ruhr, on the other hand, another class of older workers (*Zimmer- and Reparaturhauer*) was responsible for this work and they were paid a fixed wage.[55] There were also differences in the structure of the miners' working lives. In the Ruhr only youths over the age of sixteen were permitted to work underground. This rule had survived the economic liberalisation of the industry and the *Oberbergamt* resisted employers' arguments for a reduction of the age limit. Younger boys could only be employed on surface jobs. After their sixteenth birthday the youths, and older immigrants new to the work, could enter the pit as hauliers (*Schlepper*). After an unspecified length of time, which usually depended on economic conditions, the haulier could become an apprentice hewer. He would work under an experienced man to learn the skills necessary for coal getting. A full hewer had to be at least twenty-one years old and have three years' experience underground, including one as an apprentice hewer.[56]

In south Wales youths generally entered the pit at a younger age. Most seem to have had their first experience of underground work at around fourteen. The number of youths between fourteen and sixteen employed in the Welsh industry formed a substantially greater proportion of the total workforce than in the Ruhr. In 1914 this age group composed 11 per cent of the workforce in south Wales, while as a consequence of the mining regulations described above they formed only 3.7 per cent in the Ruhr in 1913.[57] Furthermore, rather than progressing through systematic stages, Welsh miners often began their working lives as either trainee hauliers or hewers. Therefore, the occupational grade of haulier was not composed of a distinctive adolescent age group as in the Ruhr. In 1906 only 3.4 per cent of hauliers were aged under sixteen, compared with

14.5 per cent of face workers in south Wales.[58] Often the apprentice and his mentor would be related to each other. In both areas the task of the haulier was considered to be inferior to that of the hewer. In tangible terms the superiority of the latter was demonstrated by higher wages. In the Ruhr the haulier earned around 60 per cent of the hewer's wage, while in south Wales it was around 69 per cent.[59]

By the 1890s three methods were used in the process of coal getting: the pillar and bord; the pillar and stall; and the longwall system.[60] In the Ruhr the former system was the dominant method of production accounting for almost half of total output in 1901, but by the 1890s both the longwall and pillar and stall systems were being increasingly used.[61] Due to the specific geology of the south Wales coalfield, employers had moved away from pillar and bord to longwall mining much earlier and by the 1890s it was already the predominant method used.[62]

Apart from the cramped, dirty, hot and often wet working conditions, mining was a dangerous occupation and south Wales and the Ruhr were no exception. Falls from the roof or sides, flooding and explosions of gas could all cause great loss of life. In March 1889 87 men were killed at an explosion at the Morfa colliery; in 1893 290 men died at the Albion mine; while 439 lost their lives in the most famous Welsh disaster at Senghennydd in 1913.[63] The Ruhr too had its major disasters. In 1898 116 died in an explosion at the *Carolinenglück*, while 114 were killed at *Lotheringen* in 1912. The worst explosion occurred at the *Radbod* mine in November 1908 where 348 workers lost their lives.[64]

These large-scale disasters, while temporarily awakening the wider world to the danger of the job, only contributed to a small proportion of those killed. Far more common were the small accidents that crippled or killed a man or two. Bert Coombes recalled several hair-raising incidents in his memoirs.[65] In 1910 a total of 1,390 accidents were reported to the Inspector of Mines. Of these 302 resulted in 313 deaths, while 1,170 men were injured.[66] The most common injuries were bone fractures, with general lacerations and contusions coming second.[67] Even seemingly trivial incidents could lead to death.[68] Falls were also a problem in the Ruhr with 45 per cent of accidents thought to be due to falls in 1885/6, decreasing to a third by 1901.[69] Similar causes were identified for the high number of accidents. Both HM Inspector of Mines and a Coal Commission established by the German Government in 1897 identified the insufficient use of props as a major danger.[70] However, these complaints largely fell on deaf ears. An attempt by the *Oberbergamt* to persuade the mine managers to introduce a more systematic approach was countered by arguments that the variable conditions underground militated against such a system and that it should be left to the hewer to set props 'where it appears necessary for security'.[71]

Chris Williams has suggested that more consideration needs to be given by historians to the nature of mine work itself. How, for example, did the miners regard issues of craftsmanship and the introduction of machinery?[72] Unfortunately, for south Wales there seems a paucity of primary evidence, although Coombes' autobiography conveys a sense that an intensification of the work process was detrimental to the miners' skills. Interestingly, for the Ruhr we are given some insight to these questions by Adolf Levenstein's 1912 survey. When asked if they enjoyed their work 507 miners claimed they did not, compared to 171 who did.[73] When asked what work they would like to do, only 68 out of 596 respondents answered mining.[74] Moreover, it appears that in south Wales mining was often a hereditary occupation. Sons followed their fathers down the pit, even when parents would have preferred them to continue their education.[75] The hereditary nature of mining in the Ruhr had been broken in the 1890s by the influx of immigrant labour.[76] This tendency was no doubt accentuated by the various other industrial jobs available in the Ruhr. Klaus Tenfelde has argued that the increasing number of officials, coupled with the introduction of longwall working, entailed a loss of autonomy for the Ruhr miners.[77] This contributed to a feeling among the miners that they were being de-skilled and led them to encourage their children into other sectors.[78] Although many Welsh miners undoubtedly also encouraged their children to seek better work, it is revealing that many miners who obtained the qualifications to become an official refused to accept such posts.[79] In south Wales then mining seems to have retained a higher status than in the Ruhr, if only due to a lack of alternatives.

There existed many possible points of conflict between the management and the workforce. One of the most obvious was the piece-rate for hewing. Miners in both areas would cite specific geological conditions of the working places, which affected the quantity of coal that could be won, when negotiating with management. In the Ruhr bargaining would be held at the start of every month between the leader of the *Kameradschaften* (*Ortältesten*) and the undermanager, while in south Wales a price list would be set for a particular seam and subject to re-negotiation for 'abnormal' places.[80] Since labour formed the greatest proportion of the cost of production in the industry it was in the management's interests to keep rates as low as possible, placing themselves and the workforce in direct opposition, offering much scope for conflict. For example, in south Wales the industrial unrest of 1910 originated in a dispute over the rate of pay for the Bute seam at Ely, part of the Cambrian Combine. Failure of the management and workforce to reach agreement led the former to post lockout notices precipitating a bitter, drawn-out struggle.[81] The Ruhr miners were also very conscious of their earnings. A report from the district of East Dortmund to the *Oberbergamt* claimed that 'the worker himself knows exactly how much work in his stall or pillar is needed to fill a wagon. Should the mine management introduce larger wagons, the worker would immediately demand a rise in his wages.' The report claimed that the workers could also judge the veins and the size of coal produced.[82]

Another related point of conflict was deductions from wages. In both areas there existed a system whereby trams deemed to contain too much small coal or rubbish resulted in deductions from the hewers' earnings. Known as 'cropping' in south Wales, this system was ultimately replaced by the use of screens that sorted the small coal from the large.[83] Furthermore, concerns that companies were underpaying for large coal led, in 1860, to the institution of checkweighers to see that the correct weight of large coal was recorded and paid for. In the Ruhr, on the other hand, the system of deductions, or *Nullen*, survived until 1905 and proved a major source of discontent.

The use of piecework and the system of deduction and coal screening can be regarded as means by which the employers could exercise some element of control over their workforce. Piecework ensured a reasonable output in a physical environment that precluded total supervision. Within this system the miners were left with a great deal of autonomy. This issue of control at work was a further potential point of conflict. It was most obviously illustrated when disputes arose over long-established customs. This lay at the root of the unrest of 1910 as, almost simultaneous with the Cambrian Combine dispute, a strike broke out in the Aberdare valley over the company's attempt to abolish the miners' customary right to take home spare pit wood.[84] Challenges could also be more direct. On 21 January 1886 colliers working on the nine-foot seam at the Plymouth colliery refused to enter the mine, citing the presence of gas as the reason, although the General Managers believed the real issue was the unpopularity of another workman.[85] A strike began despite the fact that an investigation of the mine had revealed that it was clear. A few days later Bailey recorded in his diary, 'David Morgan (the local miners' agent) came to see me … with three resolutions which he said I must comply with and this I flatly declined to do telling him that I should not allow him or anyone else to interfere with my management.'[86]

The protection of customary rights also drove the Ruhr miners to action as illustrated by the 1876 strike at the *Borussia* mine. According to the works contract the workers should have performed eight hours' work at the face. However, over time the winding time had come to be included in those hours, thereby reducing the actual time spent winning coal. The management decided to enforce the provisions of the contract and put up a poster to this effect. Finding their proclamation ignored by the workers the management adopted a more forceful approach, precipitating a strike.[87] The manager behind this strike, Grau, was involved in an almost identical dispute seven years later at the *Germania* mine. Here the previous manager, Kracher, had 'possessed the trust of the workers to the greatest degree' and his departure was 'much regretted by the whole workforce'. Grau did not command such sympathy and his attempt strictly to enforce working time again caused a strike.[88]

Strikes, however, were only one way by which the miners could register their discontent. More often miners in the Ruhr simply moved to another mine and labour turnover was exceptionally high. Roughly forty per cent of miners changed their working places every year in the Ruhr.[89] Often movement was motivated by the promise of increased wages, although conflict with the officials could be a reason. One miner's wife recalled that her father was 'a free spirit (*Wandervogel*). When he had a bust up with the undermanager he demanded his papers. That happened every once in a while, then he'd move on and, after fourteen days, write to my mother, "come".'[90]

As these examples show, miners in both areas were very protective of what they regarded as their customary rights. There was a feeling among the Ruhr miners that their occupation had lost it previous status. However, the examples given here seem to qualify the view that they lost their solidarity and autonomy as a result of the liberalisation of the industry.[91] The experience of underground work and the great potential for conflict with management generated a great degree of solidarity among the working groups. Such a view has been expressed before, but however much underground work might 'naturally' incline individuals to solidarity, there is no guarantee that such feelings retained their force after the miners had passed through the works' gates.

## The miners and their communities

The massive expansion of the coal industry had a comparable effect upon the demographics of both regions. The promise of good wages combined with the desire to escape the drudgery of farm work acted as a suitable lure for many men. Such was the attraction for Bert Coombes who sought 'good clothes, money to spend'.[92] Higher wages also attracted migrants from other industries, such as Hans Marchwitza. Having already worked underground in Upper Siliesia he left for the Ruhr after hearing that the pay was better and the hours shorter.[93] Other migrants left their home regions wanting to see more of life. J. L. Williams, for example, had a sense of a 'wider life' than on the north Wales farm where he grew up.[94]

Between 1901 and 1911 the Rhondda experienced a 34.3 per cent expansion in population, while the average increase in England and Wales was 10.9 per cent.[95] In the Ruhr the increase was even larger. It mushroomed by 75 and 102 per cent in the Rhineland and Westphalia respectively, while the average for Germany was 44 per cent.[96]

The pattern of migration into the two regions was the same. The industry first drew workers from the neighbouring areas (*Nahwanderer*), before recruiting from further afield (*Fernwanderer*). As the mining population rose so the numbers engaged in agricultural work fell. In south Wales the percentage of the labour force involved in agriculture fell from 35 per cent in 1851 to 12 per cent by 1911.[97] In the Ruhr, communities such as Steinfeld experienced a change in the traditional occupational structure as the young men entered the pit rather than taking up the agricultural work of previous generations.[98] There were differences between regions. While the Ruhr and Hellweg zones drew most of their labour from the existing population, the Emscher area was more thinly populated and, therefore, attracted the most immigrants.[99] These immigrants were often from the *Junker* estates of East Prussia. By 1907 these immigrants made up 27.3 and 44.9 per cent of the total immigrant population in the Rhineland and Westphalia respectively.[100] On a local level, in 1907 49 per cent of Essen's population had been born there,[101] while in Bochum only 23 per cent of the workforce was native to the city.[102] Many of these migrants found their way to the mines (see Table 1.5).

*Table 1.5* Percentage of mining workforce in the Ruhr born in eastern German provinces and abroad

| | 1893 | 1904 | 1912 |
|---|---|---|---|
| Eastern provinces | 25.5 | 32.2 | 36.8 |
| Foreign countries | 2.8 | 6.5 | 8.3 |

*Source:* Berg, *Wirtschaft und Gesellschaft*, p. 241.

In some areas the migrants and their descendants eventually formed a majority as villages exploded into towns. In 1893 migrants accounted for 45 per cent of the mining workforce of Bottrop. By 1920 they accounted for 75 per cent of the population. Among the migrants to the mining community of Bottrop, Poles, mainly from Posen and Upper Silesia, formed the largest single group.[103]

Similarly, while the first waves of migration to industrial south Wales came from the principality's rural counties, a trend that may have actually strengthened the Welsh language, subsequent waves came increasingly from England.[104] The number of Rhondda inhabitants who had been born in England rose from 7,328 to 10,728 between 1891 and 1911.[105] Coombes recorded the mix of dialects and languages at lodge meetings. 'Yorkshire and Durham men, Londoners, men from the Forest of Dean, North Welshmen, two Australians, four Frenchmen, and several coloured gentlemen' were represented.[106] Despite this linguistic kaleidoscope it is clear that migrants to south Wales were a more homogenous group than in the Ruhr. The vast majority of migrants entered the Glamorgan coalfield from either Wales or England, as shown in Table 1.6.

*Table 1.6* Origin of migrants to Glamorgan

| | 1871–1881 | 1881–1891 | 1891–1901 | 1901–1911 |
|---|---|---|---|---|
| England | 42,300 | 56,700 | 42,000 | 78,600 |
| Wales | 31,400 | 47,200 | 44,600 | 44,300 |
| Scotland | 1,000 | 1,600 | 1,200 | 1,800 |
| Ireland | | 3,000 | 6,600 | 3,800 |
| Total | 74,700 | 108,500 | 94,400 | 128,500 |

*Source:* Brinley Thomas, 'The Migration of Labour into the Glamorganshire Coalfield, 1861–1911', in W. E. Minchinton (ed.) *Industrial South Wales 1750–1914: Essays in Welsh Economic History* (London: Frank Cass and Company Ltd), 1968, p. 293.

Most immigrants to Glamorgan came from England. Early immigrants seem to have been assimilated into Welsh language communities, if not exactly smoothly, then without great difficulty. However, the greater number of English migrants towards the end of the nineteenth century made this more difficult.[107] This undoubtedly had an impact of the Welsh language, but it also appears that the Welsh mining communities chose English as the medium through which to conduct their business. Moreover the Welsh turned increasingly to English as a means of social mobility, since English was already established as the language of commerce, education and state bureaucracy.[108]

The industrial communities of the Ruhr, where it was believed 'some twenty languages and idioms' were spoken,[109] had a greater degree of diversity. Among the migrants the Poles stood out as the largest and most cohesive group. In 1903 Lorenz Pieper identified 19 'Polish' mines, at which more than half the work-force was made up of Polish workers. At the *Pluto* mine in Gelsenkirchen they made up almost three-quarters of the workforce.[110] While the early Polish immigrants to the area had been largely assimilated into the existing Catholic community through common religious bonds, this integration became increasingly difficult from the 1890s as the number of migrants mushroomed. Low-level discrimination by the indigenous population and the government's 'Germanisation' policy forced this second wave of immigrants to establish their own distinct subculture. A Polish language newspaper, *Warius Polski*, founded in 1890, served this milieu. Initially intended to integrate the Poles, the paper actually played a significant role in consolidating the Poles as a distinct section of the community.[111] This subculture was made of a plethora of Polish organisations ranging from educational to workers' associations.[112] Early Polish associations had a religious, Catholic character, but from the mid-1890s there was an increase in secular organisations such as the *Zwiazek Polakow w Niemczech* (Union of Poles in Germany) established in 1894. This in turn was instrumental in forming the Main Voting Committee for Westphalia, the Rhineland and Neighbouring Western Provinces in 1897. This organisation was dedicated to running – and campaigning for Polish candidates in – national and local elections. Other bodies included the *Sokol* associations, gymnastic groups modelled on the German *Turnvereine*.[113] In Bochum there were nine Polish associations with a total membership of 754 in 1893, plus another two organisations in the city itself.[114] Meanwhile, the *Regierungspräsident* (District Minister) for Arnsberg reported 191 such organisations for the area as a whole in 1906.[115] By 1912 there were 1,038 Polish associations, over a half of which were secular. The total membership of these associations was around 111,444, although this undoubtedly included dual memberships.[116]

Among these Polish organisations was the miners' union, *Zjednoczenie Zawodowe Polskie* (ZZP). Established in 1902, the founders believed that the interests of Polish workers were not being adequately protected by the existing trade unions, the *Alter Verband* (Old Association) and the *Gewerkverein christlicher Bergarbeiter* (Trade Association of Christian Miners). This view was strengthened by the old unions' support for the Mining Language Ordinance. An example of the state 'Germanisation' policy, the ordinance effectively restricted immigrants' mobility within the workplace by requiring foremen to be able to speak and write German fluently.[117] Therefore, in direct contrast to the position of English in south Wales, Polish in the Ruhr was a barrier to social mobility and few German miners' leaders felt inclined to learn the language.

Although the Poles were the largest single group, other migrants also formed subcultures. The Masurians also came from East Prussia, but were distinguished from the Poles by dialect and by their Protestantism. They were proud to be called Old Prussians and regarded any confusion of themselves with the Poles as an insult.[118] The German state deliberately fostered a Masurian identity through financial assistance for evangelical Polish-speaking pastors and a Masurian press. However, hopes that the Masurians would assimilate into wider German society were dashed as they formed their own subculture.[119] Moreover, despite these assimilatory efforts both the Poles and Masurians were often lumped together in official statistics.[120] Several smaller migrant groups, such as those from Hesse, also formed their own cultural associations.[121]

Apart from the greater ethnic diversity of the migrants to the Ruhr, the settlement of the migrants also differed. The Ruhr underwent a much greater degree of urbanisation than south Wales. Older centres like Essen and Bochum expanded rapidly, while others such as Gelsenkirchen, which only came into being in 1903, were creations of industrialisation.[122] The multitude of urban centres in the Ruhr was facilitated by Germany's history of federalism, which encouraged development of the seats of government of the various states.[123] Furthermore, owing to the greater industrial diversity of the region, the towns and cities accommodated different occupational groups. Miners, although they may have congregated in certain residential areas,[124] rubbed shoulders with metalworkers. For example, Essen was known as the *Kanonenstadt*, because of the dominance of Krupp's armament factories.[125] However, the industrial pit village also existed in the Ruhr, especially in the Emscher region. Examples of this type of settlement included communities such as Gladbeck, Buer and Herten.[126]

The south Wales valleys were prejudicial to the development of large towns and cities. Building tended to be high-density terraced housing in ribbons along the valley sides and the pit village was more common than the large conurbation.[127] The latter developed beyond the coalfield at the point of export. The Welsh mining communities were also, unlike those of the Ruhr, more geographically isolated. As the valleys were virtually mono-industrial the south Wales miners also had less opportunity to mix with other types of worker.

A further difference between the mining communities was the extent of company housing. This form of accommodation gradually decreased in south Wales towards the end of the nineteenth century. Simultaneously there was a growing trend towards home ownership facilitated by building clubs.[128] It has been estimated that these building clubs were responsible for 23 per cent of all house construction in the Rhondda in the 1890s.[129] In his memoirs Walter Haydn Davies records that his family was a member of such a club and, through their 'frugal, hardworking' lifestyle, were able to acquire two properties, one of which they rented out.[130]

Welsh employers were not insensible to the benefits of company housing. The Powell Duffryn Company recognised the 'urgent necessity which exists for providing further Cottage accommodation for the Company's workmen and the benefit which would result to the Company there from'.[131] However, such housing did not always guarantee a stable and loyal workforce.

The manager of Locket's Merthyr bemoaned the fact that 'the hold we had upon the Mardy Cottage Co.'s houses is now lost, and the result is that several tenants are now working at Ferndale, to the detriment of our Collieries'.[132] Furthermore, some thirty company houses were found abandoned in 1894 after a dispute at the South Dunraven colliery, Blaenrhondda.[133]

Welsh coalowners were, however, largely disinclined to build housing as they usually held only short leases for mineral rights.[134] The high proportion of home ownership suggests that the Welsh labour force was more stable than that in the Ruhr. Speculative builders carried out much of the rest of the house building, aided by the low costs of such construction.[135]

On the other hand, company housing (*Kolonie*) was a major feature in the Ruhr. Dick Geary has convincingly argued that such paternalistic practices were not a sign of the 'feudal' character of the German employers, but a rational response to specific problems.[136] Within the context of a particularly high turnover of labour, company housing was seen as one method of establishing a stable workforce. By 1913 there were only three coalowners without any company accommodation.[137] The Helene and Amalie mines alone owned 347 houses and 1,199 flats.[138] Often accommodating *Fernwanderer*, colonies could undermine the formation of community solidarity by isolating their residents from the other workers.[139] In fact, some housing was advertised on the basis that only certain ethnic groups would be included. A *Kolonie* attached to the *Victor* mine in Rauxel advertised itself as exclusively Masurian and inhabitants had 'nothing to do with Poles and East Prussians'.[140] Such housing was often much sought after as it was larger than accommodation available on the open market and the rent was roughly 50 per cent cheaper. As a result the proportion of miners who found lodgings in this type of accommodation grew from 4 per cent in 1872 to between 35 and 38 per cent in 1914.[141]

Alongside the *Kolonie* resident, another type of miner has been distinguished. In the older mining areas working underground was a family tradition. These families often had their origins in the pre-capitalist period. Referred to as *Kötterbergleute* (cottage miners) these miners were often smallholders, combining agricultural work with mining.[142] Heinrich Imbusch, who rose to become a leading figure in the *Gewerkverein*, came from such a background, as his family owned some livestock and even employed a housemaid.[143] *Kolonie* accommodation often included gardens and allotments by which the miners could supplement their diet and many mining families owned some form of small livestock such as pigs, sheep and goats.[144] Such livestock rearing was less common in south Wales, although in the western anthracite region, an area characterised by small mines of between 100 to 500 men, miners often engaged in both industrial and agricultural work.[145]

Relations between the older, established miners and the immigrant *Kolonie* residents varied from hostility to solidarity. The indigenous inhabitants of Steinfeld regarded the *Kolonie* inhabitants as lower-class citizens. Tensions were intensified by religious differences. Inhabitants of the *Kolonie* were largely Protestant while villagers were Catholic.[146] Meanwhile, the community of

Gladbeck exhibited a greater degree of solidarity with *Kolonists*, a feature reinforced by the shared Catholicism of both groups.

Despite the variety of accommodation there was simply not enough to cope with the massive migration into the two areas in the latter half of the nineteenth century. To some extent this problem was overcome by the system of lodging, a feature that was common to both areas. Franz Brüggemeier has argued that this led to a 'half-open family structure' among the Ruhr miners whereby lodgers were accepted into the family. An attempt to organise everyday life, it underlay the solidarity of the Ruhr miners during the strikes of 1889 and 1905.[147] In these terms the concept seems equally applicable to the south Wales coalfield. Lodging could be mutually beneficial, supplying the family with an extra source of income and the newly arrived immigrants with necessary social support. However, it could also undermine occupational solidarity in favour of ethnic or religious identities. Immigrants to the Ruhr, for example, often found lodgings with families from the same region who spoke the same language.[148] This served to strengthen ethnic identities in the region and whole urban districts could become dominated by one ethnic group.[149]

Lodgers could also mean an increased burden of work for the miner's wife. This would seem to be especially the case in south Wales because of the absence of pithead baths before the Great War. The German miners' wives were spared the drudgery of repeatedly heating water baths. But in both regions the role of women was largely restricted to the home. The dominance of heavy industry meant that there were few opportunities for paid female employment outside the home, although women could supplement a family income through some form of 'penny capitalism', such as making treats or taking in washing.[150] Outside the home both mining regions were very much masculine worlds.

So far this chapter has painted a picture of the south Wales mining communities as comparatively more homogenous then those in the Ruhr. Occupational and community solidarity was fostered by the dominance of coal mining as the male occupation coupled with the comparative isolation of the pit villages. They were not, however, immune to ethnic tensions. Many examples could be cited. In 1882 the houses of Irish workers were burnt in Tredegar.[151] Four years later extra police had to be drafted into Blaenau Gwent because of fighting between the native Welsh and English immigrants from the Forest of Dean.[152] In 1911 rioters turned on Jewish owned shops.[153] Referring to a court case involving an Italian miner, the *Porth Gazette* reported in 1902 that 'the employment of foreigners in the Rhondda coal mines has been strenuously objected to by the local colliers'.[154]

Yet, unlike the Ruhr, where the Poles felt compelled to establish their own union, there was no translation of ethnic tensions into organisational rivalry. When the issue of foreign workers was raised at a meeting of the Rhondda miners, their agent William Abraham commented 'that a man was a Frenchman or an Italian was no reason why he should not work at the pits, but if they were unskilled workmen he advised the skilled workmen not to work with them'. He claimed it was hardly fair to punish foreigners for breaking rules which they did not understand. Instead, it was the duty of the management to post the rules in a language they could understand.[155] This attitude contrasts sharply to the official support the *Alter Verband* and the *Gewerkverein* gave to the Mining Language Ordinance, which placed the onus on the immigrants to learn German.

Religion cut across these ethnic divisions. Although the actual power of the Nonconformist chapels in south Wales should not be over-estimated they did provide a form of social cement in communities which had sprung up, often from nothing, with astonishing rapidity.[156] Apart from providing spiritual sustenance they also acted as meeting places for literary societies or physical education groups. For example, Chapel Penuel in Ebbw Vale hosted meetings of the Young People's Temperance and Literary Societies ranging from simple musical evenings with recitations to discussions on topics such as 'Is ritualism consistent with the Christianity of the New Testament?'.[157] By the 1890s and 1900s the chapels had to compete with less religious entertainments. One exminer commented that prior to 1900 'the chapel was very powerful in the Rhondda Valley for the simple reason there was no other entertainment'.[158] But in other senses the chapels were quick to adapt. For example, the Temple Baptist Chapel at Graig, Pontypridd, which was established to serve the mining communities that grew up around the Maritime Collieries, changed its sermons from Welsh to English in order to minister to the large influx of English immigrants.[159]

Religion in the Ruhr was divisive. The area contained significant numbers of both Protestant and Catholic denominations although their distribution was uneven. In 1912 the city of Bochum had an almost equal proportion of both, while in the surrounding county (*Landkreis*) Protestants accounted for 61 per cent of the population, Catholics for 37 per cent. A closer look at the coal mining areas, which lay mainly in the *Landkreis*, reveal that some were dominated by Catholics (Riemke), others by Protestants (Harpen and Bergen), while some were roughly equally divided (Hordel).[160] Although the Ruhr like other industrial regions was undergoing a slow process of secularisation, religious identity remained an important defining feature among the miners. Religious divisions between Catholicism and Protestantism manifested themselves in social provisions.[161] In 1900, Bochum's four Catholic churches and eight Catholic schools were matched by four Protestant churches and seven Protestant schools. There was also a synagogue, a Jewish school and a small Baptist community.[162]

The divisions between Catholic and Protestant were not only horizontal but vertical. Catholics were over-represented in the working and lower middle classes of German society, while the coalowners of the Ruhr were largely Protestant. Thyssen was a notable exception.[163] The proportion of Catholics among the miners was increased by the arrival of the immigrants as the Poles especially tended to be Catholic. The Catholic population of Bottrop, for example, rose to 91 per cent by 1900.

This Catholic community was consolidated during the 1870s by the state-initiated *Kulturkampf* (Struggle of Cultures), which discriminated against the Catholic Church. This consolidation produced a distinctive Catholic milieu,[164] and the Ruhr, with its unique confessional structure, provided one of its strongholds.[165] The political expression of this milieu was the Catholic Centre Party.[166] However, it was also made up of a dense network of ancillary organisations. For example, in response to the liberalisation of the industry *Knappenvereine* and *Arbeitervereine* (Workers' Associations) were established. Often led by priests, these organisations sought to provide their members with sickness and burial benefits, while also maintaining the sense of status the miners had enjoyed under state control.[167] By 1897 there were 91 Catholic *Arbeitervereine* in the bishopric of Essen.[168] By 1912 Ruhr miners made up 44.4 per cent of the membership of the *Verband katholischer Arbeitervereine Westdeutschlands* (Association of Catholic West German Workers' Associations).[169] The Catholic subculture was also made up of various other religious and educational societies. This dense network was to provide an integrative function for many of the early immigrants to the region.[170]

However, shared religion was no guarantee of assimilation. For example, the Polish immigrants to Bottrop found that their spiritual needs were not well served. There was only one church and no services in the Polish language. The Poles petitioned for better provision in 1885, but it was only in 1902 that a Polish-speaking priest arrived to administer to the Herz Jesu church.[171] Such complaints became more common as the Polish communities became more established. In 1904 the District Minister for Düsseldorf saw as noteworthy 'the increasingly audacious demands for greater consideration of the Polish language in Church and School'. He also commented on complaints made to the Bishops of Münster and Paderborn that nowhere in their dioceses were there Polish services more than twice a month and that confession was exclusively in German.[172] In fact many Poles believed that they were regarded by the German Catholics as somehow inferior Catholics. At one meeting they complained they lived 'worse than a sect here' and threatened not to attend Church until a Polish priest was provided.[173]

More often the Poles established organisations to support candidates seeking election to Church boards. In Bottrop the Poles had succeeded in capturing nine out of ten seats on the Church executive, but rather than encouraging closer cooperation this seemed to heighten the mistrust between the German locals and Polish migrants.[174] Furthermore the state's Germanisation policy of the late nineteenth century, much like the *Kulturkampf* of the 1870s, served to consolidate the Polish subculture, rather than aid assimilation.[175]

Social-democratic organisations co-existed alongside these Catholic and Polish associations. Like the Catholics, the Social Democrats had been the victims of a state-led policy of repression. However, following the lapsing of the Anti-Socialist Law and the miners' strike of 1889, social-democratic party organisations, which had often disguised themselves as choral groups,[176] quickly reappeared in the Ruhr.[177] The SPD also sought to create its own network of clubs. Philip Sommerlad, a social-democratic activist in Bochum, recalled that the first cells of the local movement were made up of educational societies, while in Riemke there was a choral group, *Morgenrot*.[178] Among these clubs were the social-democratic *Konsumvereine* (Co-operatives) and *Schnapskasinos* (members' drinking clubs).[179] In 1902 the Citizens and Workers Co-operative was founded in Essen; by 1904 it had swelled to 1,724 members.[180] Meanwhile, the *Oberbergamt* estimated that there were 117 *Schnapskasinos* in its area of jurisdiction, although the definition of what constituted a *Schnapskasino* often varied.[181] Not all the *Konsumvereine* and *Schnapskasinos* were social-democratic organisations and the party's attitude to the latter was especially complicated. Often they were condemned as a distraction and a drain on wages, but they also offered a place to hold political meetings.[182]

Vernon L. Lidtke has described these associations as representing an attempt to create an alternative culture.[183] However, Karl Rohe, in applying the concept of the milieu to the Ruhr, argues that, although it is difficult to quantify the degree to which the SPD were able to build a social-democratic milieu through the foundation of associations, this construction was generally weak. While the Centre represented 'the milieu party *par excellence*', the SPD was more of a *Sammlungsbewegung* (protest movement). Yet Rohe does believe that it possessed a certain milieu foundation. Significantly, he sees it as based mainly upon the Protestant cottage miners.[184] Government officials provided some evidence to back this claim. In 1890 the District Minister for Arnsberg noted that some 46 per cent of the SPD activists in his area were, or had been, miners.[185]

This picture of milieux and subcultures can be overdrawn. The same organisations that divided the miners into certain milieu could also help them integrate into industrial society and create a broader working-class culture.[186] Moreover, an increasing commercialisation of leisure from the 1900s tended to break down barriers based on religion and ethnicity.[187] Generally, however, south Wales represents a contrast to the divisions in the Ruhr. There were, of course, countervailing forces. Nonconformity was divided into various sects, while Conservative working men's clubs jostled with classes held by former Ruskin, and later Central Labour College, students.[188] However, there were two key differences between associational life in south Wales and the Ruhr. First, the political and religious organisations in south Wales did not equal the attempts made by Catholics, Poles and Social Democrats to create a club-life that could virtually cater for an individual's every need. Secondly, in south Wales plural group membership was generally accepted, whereas in the Ruhr it was often exclusive. This difference is illustrated by the relationship between the SWMF and Labour. Although the SWMF eventually affiliated to the Labour Party, its members were free to join other political parties. On the other hand, Catholics joining the SPD risked ostracism from the Catholic milieu and even excommunication. Moreover, in south Wales both rugby and football provided a focus for the community as a whole, rather than a particular political or ethnic group.[189] Although apolitical sporting clubs, such as Schalke 04, also existed in the Ruhr, many clubs demonstrated their political loyalties by affiliating to the social-democratic *Arbeiter-Turn und Sportbund* (Workers' Gymnastic and Sports Union) or the middle-class, Protestant *Deutsche Turnschaft* (German Gymnastic Association).[190] Similarly, although Welsh employers often provided funds for workmen's halls and libraries they seem to rarely have used them as a means of social control. The committees of many libraries and halls later fell into the hands of the labour movement, but meeting rooms remained open to various causes and books ranged from non-political fiction to socialist works.[191] By contrast, the trade unions in the Ruhr built meeting halls and libraries specifically for their own members.[192] Therefore, civil society in south Wales, although pluralistic, was not so rigidly segmented by ethnicity, religion and politics. On the other hand, the establishment of distinct milieux in the Ruhr meant civil society was more rigidly divided.

## Summary

This 'snapshot' of the two coalfields has attempted to illustrate some of the long-term structures and trends within the development of the two coalfields prior to the outbreak of the Great War. How far these trends applied to other coalfields in Britain and Germany is questionable. Certainly, some developments were shared, but others were specific to certain coalfields or industrial areas. Two issues can be used to illustrate this point: ownership and the impact of immigration. For example, in Scotland, as in south Wales, there was a trend after 1870 from individually owned mines to joint-stock companies. In Germany, however, while the Ruhr mining industry was liberalised, the Saar remained under state control. Moreover, the Rhineland–Westphalian area within Prussia had certain liberal traditions, which stretched back to Napoleon and the creation of the Kingdom of Westphalia, that were not shared with other parts of Prussia, such as Silesia.

Turning to immigration, most coalfields in Britain and Germany drew upon immigrant labour at some point. It has been the contention of this chapter that the greater degree of sectarian tension in the Ruhr compared to south Wales was complicated by the arrival of large numbers of Catholic Poles. Yet, although south Wales seems to have lacked comparable tensions, they certainly existed within other areas, such as the Scottish coalfield and the Liverpool docks.

If regional specificities set south Wales and the Ruhr apart, then the model outlined here can be applied to other coalfields. Clear differences between the two regions suggest that the lifeworld of the Welsh miners was more synchronous than those in the Ruhr. To put the comparison in the context of Bulmer's eight-point ideal-type of the 'traditional mining community' outlined in the introduction, it is clear that south Wales fits that typology better than the Ruhr. In terms of economic predominance, the Ruhr developed a more diversified industrial structure and there was a strong trend towards both horizontal and vertical integration. It was more occupationally heterogeneous than south Wales, where mining was economically predominant. There the pit seemed to impinge on every aspect of life, providing a common feature in the lifeworld of the Welsh miners. Physically, it loomed over the villages. Occupationally, it was the largest employer. Organisationally, as we shall see in the next chapter, it provided the basic unit for miners' unionism. In the Ruhr, on the other hand, the pit did not dominate the lifeworld of the workers to the same extent. The winding gears shared the skyline with factory smoke-stacks, and the unions based themselves on ethnicity, religion and politics rather than the pit.

Since the nature of work underground was similar in both areas different work processes do not seem a strong enough reason for explaining dissimilarities between the two regions. However, it appears that there was to be widespread discontent with the job among the Ruhr miners. In south Wales there was greater occupational continuity between generations, if only because of lack of other work, while the 'hereditary' nature of mining was declining in the Ruhr during this period. Miners there felt that the liberalisation of the industry and the introduction of new methods of working had undermined their status. This feeling, coupled with greater opportunities for factory work and rapid expansion of the coalfield, meant that the mining workforce became increasingly made up of immigrants. Consequently, of the two regions, the Ruhr was the most ethnically diverse, with Polish migrants forming a substantial section of the labour force. In comparison migrants to south Wales were assimilated far more easily making linguistic and ethnic divisions far less evident.

Patterns of settlement also encouraged a greater synchronicity in the lifeworld of the south Wales miners. The geography of the Ruhr was more favourable for the development of cities and towns than the valleys of south Wales, which promoted the emergence of the pit village as the predominant mode of settlement. Furthermore, while a trend towards home ownership has been identified among the Welsh miners, the Ruhr saw more extensive company housing. Company housing was declining in Wales as the nineteenth century closed. In the Ruhr it increased as it offered the German industrialists a degree of social control over the mining communities that their Welsh counterparts did not enjoy. Residents of the company *Kolonie* were often isolated from the rest of the workforce and subject to strict regulations. Religion can be added to cleavages based on ethnicity and residence in the Ruhr. While Nonconformity dominated the south Wales valleys, the Ruhr miners were divided between Catholicism and Protestantism. Further complications arose when the Polish community demanded its own Catholic priests and clubs.

Divisions on ethnic, religious and political grounds influenced, and were in turn reinforced by, the various associations the miners of the Ruhr established. Pressure from the state in the form of the *Kulturkampf*, the Anti-Socialist Law and the 'Germanisation' policy served to consolidate particular groups. A distinctive Catholic milieu was formed, which was at first able to integrate the newly arrived Poles through the bond of common religion. However, from the 1890s the Poles increasingly formed their own separate associations as a reaction to a mix of discrimination and indifference. Able to act more freely after 1890, the SPD also sought to establish its own network of clubs, although the party was unable to establish the same deep roots the Catholics had.

It would not seem possible to apply the same concept of the milieu to south Wales. The theoretical literature on milieu stresses that its formation is dependent on some 'top-down' pressure. The Ruhr provides a prime example of how milieux could be formed, but similar examples are difficult to find for south Wales in the same period. As will be shown, issues such as Disestablishment, the Welsh language and Home Rule, could be effectively used to rally support. But the organisations created to pursue these aims represented the impulse to reform, rather than the defensive consolidation of an existing group in the face of state repression. Moreover, many associations, especially sporting bodies, provided a focus for the community as a whole, rather than a particular group. Lacking this 'top-down' pressure and building on a comparatively more homogeneous social base, there developed in south Wales a comparatively more diffuse, flexible civil society, which was better able to assimilate new arrivals into the existing culture than the fragmented environment of the Ruhr.

Such were the parameters within which the various miners' trade unions and political parties operated in the Ruhr and south Wales. However, it is my contention that they were not simply reflections of the structural factors set out here. Instead, they had a more dynamic role as they attempted to interpret the challenges the miners faced. It was the attempt to meet these challenges, and the solutions offered, that fed into the discursive identities of the trade unions and political parties discussed in the subsequent two chapters.

## Notes

1 Thomas Nicholas, Local Leaders Study, SWML, AUD 97.

2 L. Pieper, *Die Lage der Bergarbeiter im Ruhrrevier* (Stuttgart: J G Cotta'sche, 1903), p. 245.

3 Fagge, *Power, Culture and Conflict*, p. 15.

4 M. Atkinson, 'The supply of raw materials to the south Wales iron industry, 1800–60' and M. V. Symons, 'Coal-mining in the Llanelli area – years of growth, 1800–64', in C. Baber and L. J. Williams, *Modern South Wales: Essays in Economic History* (Cardiff: University of Wales Press, 1986), pp. 24–42 and pp. 43–52 respectively.

5 E. Weitz, 'Conflict in the Ruhr: Workers and Socialist Politics in Essen, 1910–1925' (PhD dissertation, Boston University School, 1983), pp. 32–3.

6 For the beginnings of the south Wales and the Ruhr coal industries see W. Berg, *Wirtschaft und Gesellschaft*, pp. 34–158 and E. D. Lewis, *The Rhondda Valleys: A Study in Industrial Development 1800 to the Present Day* (London: Phoenix House, 1958), p. 47.

7 W. Brephol, *Der Aufbau des Ruhrvolkes im Zuge der Ost-West-Wanderung. Beiträge zur deutschen Sozialgeschichte des 19. und 20. Jahrhunderts* (Recklinghausen: Bitter and Co, 1948), pp. 38–130.

8 R. H. Walters, *The Economic and Business History of the South Wales Coal Industry* (New York: Arno Press, 1977), pp. 49–50.

9 C. Williams, *Capitalism, Community and Conflict: The South Wales Coalfield, 1898–1947* (Cardiff: University of Wales Press, 1998), p. 89; Hickey, *Workers*, p. 13.

10 For details of the Coal Mines Act of 1842 and Coal Mines Regulation Act of 1860 see R. Church, *The History of the British Coal Industry 1830–1913: Victorian Pre-*

*eminence*, vol. 3 (Oxford: Clarendon Press, 1986), pp. 191–200.

11 Berg, *Wirtschaft und Gesellschaft*, pp. 84–8 and pp. 138–43; R. Walters, 'Capital formation in the south Wales coal industry, 1840–1914', *WHR*, 10, 1980, 69–92; Hickey, *Workers*, p. 14; G. Stolper, *The German Economy 1870–1940* (London: Allen & Unwin Ltd, 1940), pp. 44–7; Lewis, *Rhondda Valleys*, p. 47.

12 M. Kitchen, *The Political Economy of Germany 1815–1914* (London: Croom Helm, 1978), p. 95.

13 *Ibid.*, p. 96; Walters, 'Capital formation', pp. 69–92.

14 See L. J. Williams, 'The coalowners', in Smith (ed.), *People and a Proletariat*, pp. 94–113; Hickey, *Workers*, pp. 14–15.

15 James Edmund Vincent, *John Nixon: Pioneer of the Steam Coal Trade in South Wales* (London: John Murray, 1900), pp. 48–9.

16 *Ibid.*, pp. 121–5.

17 M. Asteris, 'The rise and decline of south Wales coal exports, 1870–1930', *WHR*, 13:1, 1986, 27.

18 N. J. G. Pounds, *A Historical Geography of Europe, 1800–1914* (Cambridge: Cambridge University Press, 1985), p. 45 and pp. 439–40.

19 Asteris, 'Rise and decline of south Wales exports', p. 25. The Barry docks were built in 1889 and 1898 when the Marquis demanded an increased levy on coal tipped at the existing docks and a new one to be built.

20 Asteris, 'Rise and decline of south Wales exports', p. 27

21 Berg, *Wirtschaft und Gesellschaft*, p. 89.

22 *Ibid.*, p. 90.

23 Pounds, *Historical Geography*, pp. 375–6.

24 Weitz, 'Conflict in the Ruhr', pp. 32–5.

25 Williams, *Capitalism, Community and Conflict*, p. 32.

26 T. Boyns, 'Growth in the coal industry: the cases of Powell Duffryn and the Ocean Coal Company, 1864–1913' in Baber and Williams (eds), *Modern South Wales*, pp. 153–170.

27 Hickey, *Workers*, p. 15.

28 Kitchen, *Political Economy*, pp. 134–5.

29 Berg, *Wirtschaft und Gesellschaft*, p. 92.

30 Brüggemeier, *Leben vor Ort*, p. 85.

31 Berg, *Wirtschaft und Gesellschaft*, p. 94.

32 Brüggemeier, *Leben vor Ort*, p. 84.

33 See P. Osthold, *Die Geschichte des Zechenverbandes 1908–1913. Ein Beitrag zur deutschen Sozialgeschichte* (Berlin: Elsner, 1934); Stefan Prizgoda, *Unternehmerverbände im Ruhrbergbau. Zur Geschichte von Bergbau-Verein und Zechenverband 1858–1933* (Bochum: Deutsche Bergbau-Museums, 2002), p. 96.

34 Church, *History of the British Coal Industry*, pp. 665–6.

35 *Ibid.*, pp. 234–6.

36 R. P. Arnot, *South Wales Miners, Glowyr de Cymru: A History of the South Wales Miners' Federation, 1898–1914*, vol. 1, (London: Allen & Unwin, 1967), p. 42.

37 Williams, 'The coalowners', pp. 104–5. Furthermore he suggests that the formation of the MSWCOA was problematic. See chapter 6, 'The south Wales coal owners, 1873–80: Problems of unity', in L. J. Williams, *Was Wales Industrialised? Essays in modern Welsh History* (Llandysul: Gomer Press, 1995), pp. 123–50.

38 See Williams, 'The coalowners'.

39 B. Weisbrod, 'Entrepreneurial politics and industrial relations in mining in the

Ruhr region: From managerial absolutism to co-determination', in G. D. Feldman and K. Tenfelde (eds), *Workers, Owners and Politics in Coal Mining: An International Comparison of Industrial Relations* (Oxford: Berg, 1990), p. 131.
40 Kitchen, *Political Economy*, pp. 135–7.
41 O. Faust, *Vom Bremsjungen zum Betribesinspektor. Ein Leben im Ruhrbergbau 1867–1914* (Essen: Klartext, 1989), p. 194.
42 HM Inspector of Mines, South Wales District, 1891, p. 4, GRO D/D NCB 67/2/7. The south Wales District comprised the whole of Carmarthenshire, Pembrokeshire, most of Glamorgan and part of Brecknockshire.
43 Berg, *Wirtschaft und Gesellschaft*, p. 226.
44 *Ibid.*, pp. 226–8.
45 *Ibid.*, pp. 162–3.
46 HM Inspector of Mines, Swansea District, 1909, GRO D/D NCB 67/2/15, p. 57.
47 Berg, *Wirtschaft und Gesellschaft*, p. 49.
48 M. J. Daunton, 'Down the pit: Work in the Great Northern and south Wales Coalfields, 1870–1914', *EHR*, 34:4 (1981), 583; Hickey, *Workers*, p. 112; Berg, *Wirtschaft und Gesellschaft*, p. 49.
49 U. Burghardt, *Die Mechanisierung des Ruhrbergbaues 1890–1930* (Munich: Beck), 1995, p. 61.
50 Walters, *Economic and Business History*, p. 206.
51 B. L. Coombes, *These Poor Hands: The Autobiography of a Miner Working in South Wales* (London: Victor Gollancz Ltd, 1939), p. 8 and p. 103.
52 See *Glückauf*, 5 July 1902, pp. 633–44; *Glückauf*, 25 March 1911, p. 455.
53 Berg, *Wirtschaft und Gesellschaft*, p. 163.
54 See H. S. Jevons, *The British Coal Trade* (London: Kegan Paul Trench Trubner and Co. Ltd, 1915), pp. 608–17 and pp. 209–11; Brüggemeier, *Leben vor Ort*, pp. 92–4; Berg, *Wirtschaft und Gesellschaft*, pp. 218–21 and Daunton, 'Down the pit', pp. 578–97.
55 Jevons, *British Coal Trade*, pp. 611–12; Brüggemeier, *Leben vor Ort*, p. 93.
56 See Brüggemeier, *Leben vor Ort*, pp. 96–102 and Hickey, *Workers*, pp. 109–15.
57 Berg, *Wirtschaft und Gesellschaft*, p. 537.
58 Daunton, 'Down the pit', p. 590.
59 See Brüggemeier, *Leben vor Ort*, p. 99 and Berg, *Wirtschaft und Gesellschaft*, p. 545.
60 Jevons, *British Coal Trade*, pp. 203–8; Daunton, 'Down the pit', pp. 580–1.
61 Brüggemeier, *Leben vor Ort*, pp. 105–12.
62 Vincent, *John Nixon*, pp. 182–3.
63 See Reports of HM Inspector of Mines, South Wales District, 1890, p. 6, and 1894, p. 5, GRO D/D NCB 67/2/15.
64 Hickey, *Workers*, p. 118.
65 See B. Jones and C. Williams (eds), *With Dust Still in His Throat: A. B. L. Coombes Anthology* (Cardiff: University of Wales Press, 1999), pp. 44–51.
66 Reports of HM Inspector of Mines, South Wales District, 1911, p. 5, GRO D/D NCB 67/2/16.
67 *Ibid.*, 1895, p. 6.
68 The Mines Inspector for the South Western District reported the case of a man who received a seemingly innocuous cut from a falling stone only to die later from blood poisoning. Report of HM Inspector of Mines, South Western District, 1888, p. 7, GRO D/D NCB 67/2/1.
69 Hickey, *Workers*, p. 119.

70 See Reports of HM Inspector of Mines, SW District, 1888 and 1892, p. 7 and p. 11 respectively, GRO D/D NCB 67/2/1 and D/D NCB 67/2/9; Stein- und Kohlenunfallkommission, 1897, p. 197.
71 Directors of the Harpener Bergbau AG, 21 September 1903 quoted in Brüggemeier, *Leben vor Ort*, p. 120.
72 C. Williams, 'The South Wales Miners' Federation', *Llafur*, 5:3 (1990), 53–4.
73 A. Levenstein, *Die Arbeiterfrage. Mit besonderer Berücksichtigung der sozialpsychologischen Seite des modernen Grossbetriebes und der psycho-physischen Einwirkungen auf die Arbeiter* (Munich: Ernst Rheinhardt, 1912), p. 59. Levenstein collated information from questionnaires received from miners, metalworkers and textile operatives. All respondents were politically active or Free Trade Union members. Of the miners, respondents from the Ruhr formed the largest single group, numbering 810.
74 *Ibid.*, p. 161.
75 A. Burge, 'Miners' learning in the south Wales coalfield, 1900–1947', *Llafur*, 8:1 (2000), 78.
76 Pieper, *Lage der Bergarbeiter*, p. 14.
77 Tenfelde, 'Arbeitsplatz', in Conze and Ulrich Engelhardt (eds), *Arbeiter im Industrialisierungsprozeß*, pp. 328–31.
78 I. Steinisch and K. Tenfelde, 'Technischer Wandel und soziale Anpassung in der deutschen Schwerindustrie während des 19. und 20. Jahrhunderts', *AfSG*, 28 (1988), 43.
79 Burge, 'Miners' learning', pp. 90–1.
80 Berg, *Wirtschaft und Gesellschaft*, p. 220.
81 Arnot, *South Wales Miners*, vol. 1, pp. 176–267.
82 *Bergamt* East Dortmund to the *Oberbergamt*, 12 March 1884, STAM 1392.
83 Vincent, *John Nixon*, pp. 189–96.
84 Francis and Smith, *The Fed*, p. 13.
85 Diaries of Mr T. H. Bailey, 21 January 1886, GRO D/D NCB 16/37.
86 *Ibid.*, 25 January 1886, GRO D/D NCB 16/37.
87 *Bergamt* West Dortmund to *Oberbergamt*, 12 April 1876, STAM 1778.
88 *Ibid.*, 21 December 1883, STAM 1390.
89 Brüggemeier, *Leben vor Ort*, p. 123.
90 Cited in H. Croon and K. Utermann, *Zeche und Gemeinde. Untersuchungen über den Strukturwandel einer Zechengemeinde im nördlichen Ruhrgebiet* (Tübingen: J.C.B. Moheer, 1958), p. 20.
91 See Tenfelde, *Sozialgeschichte der Bergarbeiterschaft*.
92 See Coombes, *Poor Hands*, p. 8.
93 Quoted in Brüggemeier, *Leben vor Ort*, p. 18.
94 J. L. Williams interviewed by David Egan, 'South Wales coalfield history project', SWML, AUD 396, p. 3.
95 Lewis, *Rhondda Valleys*, pp. 231–2.
96 Hickey, *Workers*, p. 22.
97 J. Williams, 'The move from the land', in Trevor Herbert and Gareth Elwyn Jones (eds), *Wales 1880–1914* (Cardiff: University of Wales Press, 1988), p. 24.
98 Croon and Utermann, *Zeche und Gemeinde*, p. 28.
99 Brephol, *Aufbau des Ruhrvolkes*, p. 80.
100 Pounds, *Historical Geography*, p. 84.
101 Weitz, 'Conflict in the Ruhr', p. 47.

102 Crew, *Town in the Ruhr*, p. 60.
103 See Jäger, *Bergarbeitermilieus und Parteien*, p. 86 and R. C. Murphy, 'Polnische Bergarbeiter im Ruhrgebiet: Das Beispiel Bottrop', in H. Mommsen and U. Borsdorf (eds), *Glück Auf*, pp. 91–2.
104 B. Thomas, 'The migration of labour into the Glamorganshire coalfield, 1861–1911', in W. E. Minchinton, *Industrial South Wales* (London: Frank Cass and Co., 1969), pp. 81–99.
105 Lewis, *Rhondda Valleys*, pp. 238–9.
106 Coombes, *Poor Hands*, p. 88.
107 See B. Thomas, 'Wales and the Atlantic economy', *SJPE*, 6:3 (1959), 169–92; B. Thomas, 'A cauldron of rebirth: Population and the Welsh language in the nineteenth century', in Geraint H. Jenkins (ed.), *A Social History of the Welsh Language: The Welsh Language and its Social Domains 1801–1911* (Cardiff: University of Wales Press, 2000).
108 T. Williams, 'Language, religion, culture', in Herbert and Jones (eds), *Wales 1880–1914*, pp. 78–9; T. Williams, 'The anglicisation of south Wales', in R. Samuel (ed.), *Patriotism*, vol. 2 (London: Routledge, 1989), pp. 193–203. See also the Commission of Inquiry into Industrial Unrest 1917, p. 15.
109 Hickey, *Workers*, p. 33.
110 Pieper, *Lage der Bergarbeiter*, p. 20.
111 C. Kleßmann, *Polnische Bergarbeiter im Ruhrgebiet 1870–1945* (Göttingen: Vandenhoeck & Ruprecht, 1978), p. 58. See also C. Kleßmann, 'Polnische Bergarbeiter im Ruhrgebiet: Soziale Lage und gewerkschaftliche Organisation', in Mommsen and Borsdorf (eds), *Glück Auf*, pp. 111–30.
112 Kleßmann, *Polnische Bergarbeiter im Ruhrgebiet*, pp. 96–138.
113 B. J. McCook, 'The Borders of Integration: Polish Migrant Workers in the Ruhr Valley of Germany and the Pennsylvania Anthracite Regions of the United States, 1870–1924' (PhD dissertation, University of California, Berkeley, 2004), pp. 400–15.
114 *Stadt* Bochum to *Regierungspräsident* Arnsberg, 18 October 1893; *Landrat* Bochum to *Regierungspräsident* Arnsberg, 19 October 1893, STAM, 14045.
115 *Regierungspräsident* Arnsberg to Minister of the Interior, 10 December 1906, STAD, 9041. The *Regierungspräsident* was an unelected government-appointed official for a given administrative district. The greatest part of the Ruhr coalfield was in the administrative districts of Arnsberg, Düsseldorf and Münster.
116 McCook, 'Borders of Integration', p. 411.
117 J. J. Kulczycki, 'A trade union for Polish miners in the Ruhr: Alter Verband, Gewerkverein and Zjednoczenie Zawodowe Polskie', in K. Tenfelde (ed.), *Towards a Social History of Mining in the 19th and 20th Centuries: Papers presented to the International Mining History Congress Bochum* (Munich: Beck, 1989), pp. 610–17; Kleßmann, *Polnische Bergarbeiter*, pp. 118–19.
118 *Bezirkspolizeikommissar* Dortmund to *Regierungspräsident* Arnsberg, 23 February 1908, STAD 16019.
119 McCook, 'Borders of Integration', pp. 370–1.
120 C. Schmidt, '"Dieses einst so verlockende Ruhrland". Migranten und Migrantenkultur der Polen und Masuren in Gelsenkirchen vor dem Ersten Weltkrieg', in S. Goch and L. Heidemann (eds), *100 Jahre Bismarck. Ein Stadtteil 'mit besonderem Erneuerungsbedarf'* (Essen: Klartext, 2001), pp. 55–66.
121 A. Krus-Bonazza, '"Da waren die Hessen, die Polen und noch eine Sorte ..."

(Arbeits)-Einwanderer in Bochum-Dahlhausen zur Zeit des Kaiserreichs und der Weimarer Republik', in P. Friedemann and G. Seebold (eds), *Struktureller Wandel und kulturelles Leben Politische Kultur in Bochum 1860–1990* (Essen: Klartext, 1992), pp. 178–88.

122 W. Köllmann, 'The process of urbanisation in Germany at the height of the industrialisation period', *JCH*, 4:3 (1969), 63–4.

123 Pounds, *Historical Geography*, p. 169.

124 See, for example, Pieper's examination of the Ueckendorf community in Gelsenkirchen, Pieper, *Lage der Bergarbeiter*, pp. 252–66.

125 J. Paul, *Alfred Krupp und die Arbeiterbewegung* (Düsseldorf: Schwamm, 1987).

126 See F. Bajhor, '"Kampf ums altes Recht?" Bergbau und Bergarbeiterschaft im Industriedorf Gladbeck 1871–1889', in K. Ditt and D. Kift (eds), *1889: Bergarbeiterstreik und Wilhelminische Gesellschaft* (Hagen: Westfälisches Industriemuseum, 1989), pp. 33–53, and W. Hoffman, 'Politik in der Provinz: Kommunale Politisierung dargestellt an den sechs Industriedörfen im Landkreis Recklinghausen vom Beginn der Industrialisierung bis 1914', *Bochumer Historische Studien, Neuere Geschichte*, 16, 1996.

127 P. N. Jones, *Colliery Settlement in the South Wales Coalfield, 1850–1926* (Hull: University of Hull Publications, 1969).

128 Gilbert, *Class*, pp. 67–8; M. Daunton 'Miners' houses: South Wales and the Great Northern coalfield, 1880–1914', *IRSH*, 25 (1980), 143–75.

129 M. J. Fisk, *Housing in the Rhondda, 1800–1940* (Cardiff: Merton Priory Press Ltd, 1996), p. 55.

130 W. H. Davies, *The Right Place, The Right Time: Memories of Boyhood Days in a Welsh Mining Community* (Swansea: Christopher Davies, 1975), pp. 66–7, GRO D/D NCB16/35.

131 Minutes of the Meeting of the Board of Directors of the Powell Duffryn Steam Coal Co., 13 May 1890, GRO D/D PD 1/3.

132 Manager of Lockets Merthyr Collieries to Directors, 9 July 1909, GRO D/D NCB16/35.

133 Fisk, *Housing in the Rhondda*, p. 47

134 *Ibid.*, p. 39.

135 Berg, *Wirtschaft und Gesellschaft*, p. 562.

136 D. Geary, 'The industrial bourgeoisie and labour relations in Germany 1871–1933', in D. Blackbourn and R. J. Evans (eds), *The German Bourgeoisie: Essays on the Social History of the German Middle Class from the Late Eighteenth to the Early Twentieth Century* (London: Routledge, 1991), pp. 142–4.

137 Berg, *Wirtschaft und Gesellschaft*, p. 273.

138 *Werksgeschichte – Gewerkschaft des Steinkohlenbergwerks vereinigte Helene und Amalie während des Krieges und der Revolution bis zu Reichsmark, 1914–1924*, 1924/25, p. 13, BBA, Bestand 20, 565.

139 S. Hickey, 'The shaping of the German labour movement: Miners in the Ruhr', in R. J. Evans (ed.), *Society and Politics in Wilhelmine Germany* (London: Croom Helm, 1978), p. 217.

140 Cited in Brüggemeier, *Leben vor Ort*, p. 27.

141 *Ibid.*, pp. 50–1; Berg, *Wirtschaft und Gesellschaft*, p. 272.

142 Jäger, *Bergarbeitermilieus und Parteien*, pp. 99–100.

143 M. Schäfer, *Heinrich Imbusch. Christlicher Gewerkschaftsführer und Widerstandskämpfer* (Munich: Beck, 1990), pp. 11–12.

144 Between 1893 and 1914 the number of pigs and sheep kept by miners increased by 168 per cent and 303 per cent respectively: Berg, *Wirtschaft und Gesellschaft*, p. 268.
145 See I. Matthews, 'The world of the anthracite miner', *Llafur*, 6:1 (1992), 96–7, and J. Griffiths 'The miners' union in the anthracite coalfield', in G. A. Hughes (ed.) *Men of No Property: Studies of Welsh Trade Unionism* (Caerswys: Gwasg Gwenffrwd, 1971), pp. 60–7.
146 Croon and Utermann, *Zeche und Gemeinde*, pp. 22–4.
147 Brüggemeier, *Leben vor Ort*, pp. 62–4.
148 *Ibid.*, p. 65.
149 See Murphy, 'Polnische Bergarbeiter im Ruhrgebiet', p. 102.
150 See D. Beddoe, *Out of the Shadows: A History of Women in Twentieth-Century Wales* (Cardiff: University of Wales Press, 2000), p. 18.
151 J. Parry, 'The Tredegar anti-Irish riots of 1882', *Llafur*, 3:4 (1983), 20–3.
152 *SWDN*, 29 July 1886.
153 G. Alderman, 'The anti-Jewish riots of August 1911 in south Wales', *Llafur*, 6:2 (1972), 190–200.
154 *PG*, 12 July 1902.
155 *RL*, 27 October 1900.
156 T. Williams estimates that about 75 per cent of the Welsh population visited the chapels between 1880 and 1914. However, he qualifies this by pointing out that the chapels often had to compete for working-class attention with more secular pursuits, such as sport, and the chapels had no formal power of enforcement. Williams, 'Language', pp. 80–3.
157 Revd. E. Price, *The History of Penuel: Calvinistic Methodist Church Ebbw Vale* (Wrexham: Hughes and Son, 1925), pp. 84–94.
158 Rees Davies, SWML, AUD 207.
159 P. Elwood, *A History of Temple Chapel*, 1987, GRO D/D Bap 44/1.
160 C. Schwaderer, 'Die Reichstagswahlen und die Wahlen zur Nationalversammlung im Landkreis und der Stadt Bochum 1912–1920' (Hausarbeit der ersten Staatsprüfung für das Lehramt an Gymnasien: Bochum, 1980), p. 67 and pp. 108–24.
161 Hickey, *Workers*, pp. 70–80.
162 M. Seippel, *Bochum einst und jetzt. Ein Rück- und Rundblick bei der Wende des Jahrhunderts* (Bochum: Schürmann und Klagges, 1901/1991), pp. 60–82.
163 *Ibid.*, p. 449; E. G. Spencer, *Management and Labor in Imperial Germany: Industrialists as Employers, 1896–1914* (New Jersey: Rutgers University Press, 1984), pp. 151–6.
164 R. Lepsius, 'Parteiensystem und Sozialstruktur. Zum Problem der Demokratisierung der deutschen Gesellschaft', G. A. Ritter (ed.), *Die deutschen Parteien vor 1918* (Cologne: Kiepenhauer & Witsch, 1973), pp. 56–80; Claggett *et al.*, 'Leadership', pp. 643–64.
165 See Jäger, *Bergarbeitermilieus und Parteien.*
166 See E. L. Evans, *The German Center Party, 1870–1933: A Study in Political Catholicism* (Carbondale and Edwardsville: Southern Illinois University Press, 1981).
167 E. Kroker and W. Kroker, *Solidarität aus Tradition. Die Knappenvereine im Ruhrgebiet* (Munich: Beck, 1988), pp. 11–47.
168 H. Budde, *Christliche-Sozial an der Ruhr. Eine Volksbewegung im Zentrum der Industrie. Dokumentation eines 150jährigen Wegen* (Cologne: Orthen Druck, 1988),

pp. 65–7.

169 C. Haffert, *Die katholishen Arbeitervereine Westdeutschlands der Weimarer Republik* (Essen: Klartext, 1994), p. 244.

170 M. Schäfer, 'Das Milieu der katholischen Arbeiter im Ruhrgebiet (1890–1914)', in D. Kift (ed.), *Kirmes-Kneipe-Kino. Arbeiterkultur im Ruhrgebiet zwischen Kommerz und Kontrolle (1850–1914)* (Paderborn: Ferdinand Schöningh, 1992), pp. 198–9.

171 Murphy, 'Polnische Bergarbeiter im Ruhrgebiet', pp. 98–9.

172 *Regierungspräsident* Düsseldorf to Minister of the Interior, 15 November 1904, STAD, 9041

173 Report on a Polish meeting held on the 15 January 1905, *Oberbürgermeister* Essen to *Regierungspräsident* Düsseldorf, 18 January 1905, STAD 9056.

174 Murphy, 'Polnische Bergarbeiter im Ruhrgebiet', pp. 98–9.

175 J. J. Kulczycki, *The Foreign Worker and the German Labour Movement: Xenophobia and Solidarity in the Coal Fields of the Ruhr, 1871–1914* (Oxford: Berg, 1994) pp. 38–40.

176 A. Hellkuhl, *Empor zum Licht. Arbeitergesangvereine im westfälichen Ruhrgebiet 1878–1914* (Stuttgart and Witten: Bertold Marohl Musikverlag, 1983), pp. 29–33.

177 See S. Goch, *Arbeiterbewegung und Arbeiterkultur im Ruhrgebiet. Eine Untersuchung am Beispiel Gelsenkirchen, 1848–197* (Düsseldorf: Dorste Verlag, 1990), pp. 81–146; R. Lützenkirchen, *Der sozialdemokratische Verein für den Reichstagswahlkkreis Dortmund-Hörde. Ein Beitrage zur Parteigeschichte* (Dortmund: Historischen Vereins Dortmund, 1970).

178 P. Sommerlad, *Vom 'Hessebud' zum 'Bochumer Jungen'. 70 Jahre sozialistische Bewegung in Bochum (1878–1948)*, ed. P. Friedemann and E. Heinevetter (Essen: Klartext, 1990), p. 50.

179 G. Huck, 'Arbeiterkonsumverein und Verbraucherorganisation. Die Entwicklung der Konsumgenossenschaften im Ruhrgebiet 1860–1914', in Reulecke and Weber (eds), *Fabrik, Familie, Feierabend*, pp. 215–46; R. Joerss, 'Moralisch Verkommen und politisch Gefährlich? Die Schnapskasinos im Ruhrgebiet am Ende des 19. Jahrhunderts', *Beiträge zur Geschichte Dortmunds und der Grafschaft Mark*, vol. 92/93 (2001/2002), pp. 133–74; F. J. Brüggemeier and L. Niethammer, 'Schlafgänger, Schnapskasinos und Schwerindustrielle Kolonie. Aspekte der Arbeiterwohnungsfrage im Ruhrgebiet vor dem Ersten Weltkrieg', in Reulecke and Weber (eds), *Fabrik, Familie, Feierabend*, pp. 135–76.

180 *Regierungspräsident* Düsseldorf to Minister of the Interior, 15 November 1904, STAD 9041.

181 *Oberbergamt* to *Regierungspräsidenten* Arnsberg, Düsseldorf and Münster, 11August 1894, STAM 1834.

182 Joerss, 'Verkommen', pp. 159–63.

183 V. L. Lidtke, *The Alternative Culture: Socialist Labor in Imperial Germany* (New York and Oxford: Oxford University Press, 1985).

184 K. Rohe, 'Die Ruhrgebietssozialdemokratie im Wilhelminischen Kaiserreich und ihr politischer und kultureller Kontext', in G. A. Ritter, *Der Aufstieg der deutschen Arbeiterbewegung. Sozialdemokratie und Freie Gewerkschaften im Parteiensystem und Sozialmilieu des Kaiserreichs* (Munich: R. Oldenburg, 1990), pp. 328–44.

185 Cited in A. Herzig, 'Organisationsprobleme der Sozialdemokratie im Ruhrgebiet im 19. Jahrhundert', in L. Heid and J. H. Schoeps (eds), *Arbeit und Alltag im Revier. Arbeiterbewegung und Arbeiterkultur im westlichen Ruhrgebiet im Kaiserreich und in der Weimarer Republik* (Duisburg: Walter Braun Verlag, 1985), p. 19.

186 L. Abrams, 'Zur Entwicklung einer kommerziellen Arbeiterkultur im Ruhrgebiet (1850–1914)', in D. Kift, *Kirmes – Kneipe – Kino. Arbeiterkultur im Ruhrgebiet zwischen Kommerz und Kontrolle, 1850–1914* (Paderborn: Ferdinand Schöningh, 1992), pp. 47–9.

187 L. Abrams, *Workers' Culture in Imperial Germany: Leisure and Recreation in the Rhineland and Westphalia* (London and New York: Routledge, 1992), p. 171.

188 On the Conservative working men's clubs see F. J. Harries, *History of Conservatism in the Rhondda* (Pontypridd: Glamorgan County Times Office, 1912). On independent working-class education see R. Lewis, *Leaders and Teachers: Adult Education and the Challenge of Labour in South Wales, 1906–1940* (Cardiff: University of Wales Press, 1993), pp. 49–97.

189 See G. Williams, *1905 and All That* (Llandysul Gomer Press, 1991), p. 17 and Martin Johnes, 'That Other Game: A Social History of Soccer in South Wales, c1906–1939' (PhD dissertation, University of Wales, 1998), p. 299.

190 See S. Gehrmann, 'Fußball in einer Industrieregion. Das Beispiel F. C. Schalke 04', in Reulecke and Weber (eds), *Fabrik Famile, Feierabend*, pp. 377–98; S. Gehrmann, *Fußball, Vereine, Politik. Zur Sportgeschichte des Reviers, 1900–1940* (Essen: Reimar Hobbing, 1988).

191 See C. M. Baggs, '"Well Done, Cymmer Workmen!" The Cymmer Collieries Workmen's Library, 1893–1920', *Llafur*, 5:3 (1990), 20–5; J. Rose, *The Intellectual Life of the British Working Classes* (New Haven and London: Yale University Press, 2001), pp. 237–55.

192 See H. Imbusch, *25 Jahre Gewerkverein christlicher Bergarbeiter* (Essen: Verlag des Gewerkvereins christlicher Bergarbeiter Deutschlands, 1919).

# 2

# Trade union discourses, 1890–1914

But you can't be in theirs and ours at the same time!

Schneider, *Alter Verband* member, to Hans Marchwitza on his dual membership of the *Alter Verband* and the *Gewerkverein.*[1]

It was both an industrial and a social institution ... It differed from the normal functions of trade unions because of its more intimate involvement in the domestic and social life of the people.[2]

Will Paynter on the SWMF.

## Introduction

This chapter examines how the trade unions sought to construct miners' identities through discourse. Between 1889 and 1902, the miners in both regions were successful in forming trade unions that were to survive up to the Great War and beyond. Yet despite a similar trend towards successful combination, it soon became apparent that patterns of trade unionism differed in the two regions. While the South Wales Miners' Federation (SWMF) assumed a hegemonic position in terms of miners' organisation before the Great War, the movement in the Ruhr was progressively split along religious, political and ethnic lines. So while trade unionism in south Wales was characterised by homogenisation, in the Ruhr the overall impression is of fragmentation.

This chapter seeks to shed some light on the reasons for these divergent trends. The focus will be upon the discursive practices of the trade unions. As will become apparent, three strands – Christian-socialism, socialism (or social democracy) and syndicalism – played important roles in the discourses of the trade unions covered here. The intention here is not totally to disassociate discourse from material structures. Instead, incidences of unemployment, wage reduction or intensification of work practices were all subject to interpretation by trade union activists. These interpretative frameworks then formed the basis of both the discursive practices of the trade unions and their actions. At the same time, actions were conditioned by the unions' own organisational resources. However, the way in which they conceived and developed their own discursive identities was to have an important impact upon their success in organising the workforce.

## Ideological conflicts: Christian-socials and Social Democrats, 'Federationists' and 'Scalites'

Trade unionism in the south Wales and the Ruhr coalfields predates the 1890s, but before then attempts at organisation had proved short lived. For example, William Abraham, the most prominent labour leader in the south Wales coalfield in the late nineteenth century, earned his trade union spurs as part of the Amalgamated Association of Miners (AAM), which had some success in the region in the early 1870s, before falling foul of unfavourable economic conditions and defeat in a large strike in 1875.[3] In the Ruhr, organisation came somewhat later. The *Knappenvereine* aimed at salvaging something of the miners' privileged status after the liberalisation of the industry by providing benefits for their members. In the 1870s there were two attempts to establish trade union organisations in the Ruhr. The first followed the 1872 Essen strike and took the form of the *Grubenarbeiterverband*; the second, the *Rosenkranzverband*, came six years later. As an augur of subsequent events, the latter organisation foundered on differences between the Social Democrats and the Christian-social miners.[4]

The 1880s represented a lull in organisational activities. However, organisation was not at a complete standstill. The *Knappenvereine* reached their high point in this decade. In south Wales, the establishment of a Sliding Scale system linked wages directly to prices. Price fluctuations were to be followed by a corresponding rise or reduction in wages.[5] Organisation of a sort survived in the form of local unions, such as the Cambrian Miners' Association.[6] However, since the Scale seemingly obviated the need for genuine negotiation over wages, these associations have been termed 'quasi'-unions, whose principal function was to raise the funds necessary to support the miners' representatives on the Sliding Scale Committee.[7]

These organisational efforts formed the background to the eventual establishment of permanent trade unions in the 1890s. What is striking about the Ruhr is the role Christian thought played in organisations. First, it provided legitimacy for the activists' efforts at organisation. Second, it formed an interpretative framework through which to criticise the employers. Pope Leo XIII's *Rerum Novarum* 'On the Rights and Duties of Capital and Labour' and the Bishop of Mainz's pamphlet entitled 'The Worker Question and Christianity' gave the movement an intellectual and moral basis.[8] Furthermore the early *Knappenvereine* were predominantly priest-led, Catholic organisations. Their statutes often called upon a member to be a 'genuine Christian' and lead a life of propriety.[9]

Christian-social thought was spurred to further action by the 1889 strike. During the 1880s discontent over wages and conditions had been simmering. In

March and April 1889 meetings in Essen had formulated a list of demands regarding wages, the introduction of the eight-hour shift and restriction of overtime. The strike itself, however, was a spontaneous movement. Beginning at the *Präsident* mine in Bochum on 25 April, it quickly spread to other areas. A central strike committee was eventually established on 10 May and, five days later, a delegation presented the miners' case to the *Kaiser*. Without a strike fund and in the face of the *Bergbau Verein*'s intransigent attitude, the miners had already begun to drift back to work. On 19 May the strike was officially over.[10]

The Catholic press regarded the strike as the result of the lack of dialogue between the owners and the workers. *Tremonia*, edited by Lambert Lensing, emphatically rejected claims the strike was caused by a mixture of laziness and social-democratic agitation. Instead, it argued that the employers had a false image of the worker as a 'human machine' that needed only to work, eat and sleep. Religion was at the centre of the Christian-social critique. If 'one has the religious requirement to visit the Church on Sunday and Monday, that is already, in the eyes of many employers, a disagreeable, disruptive luxury, which hinders the organisation of the works and the worker in his work'.[11] The implication was that if the employers accepted that their workers required leisure time to fulfil their spiritual needs, then a better understanding would develop between them. Lensing further set out this conviction that religion was the key to mediating conflict between employers and employees in a pamphlet. He claimed:

> only on the basis of Christianity will the social question be solved or not ... Employer and worker must be led to a better relationship by the principle of Christian charity ... The employer should not be the tyrant of the workers, who exploits their strength like a machine, he should rather appear as a caring owner (*Fürsorger*) and regard the workers as human and equal partners, who are sooner approachable through a friendly word, than crude reprimands ... The worker on the other hand should consider the employer as his provider (*Brodherrn*), respect him, he should joyfully do his duty, whilst not stealing the day from the Lord (*wegstehlen*).[12]

Mediation of conflict between capital and labour would only be achieved through the establishment of this dialogue. Following the 1889 strike the employers attempted to introduce a system that would prevent the workers from changing workplace (*Sperre*). Along with the fines that accompanied the end of the strike, this action caused great discontent. A further strike was only avoided by the intervention of District President Berlepsch, who held talks with both miners and owners, finally persuading the latter to drop the system.[13] *Tremonia* seized upon this as a positive example of better communication.

> It has been repeatedly emphasised by experts, that with respect to an excited workforce, a friendly word is often more effective than factual concessions from the employers. That the workforce is a factor with which one must reckon has actually been recognised by the employers in their sacrifice of the *Sperre*. Why do they uphold the empty appearance of inflexibility, when they have nothing to lose, but much to gain, through the acceptance of direct negotiations?[14]

The ultimate goal of the Christian-social movement was recognition of the workers as partners in the mining industry. Strikes like that of May 1889 could be avoided through communication predicated on a shared understanding of Christian values and recognition of the importance of both employers and miners to the running of the industry.

The role of Christian thought in south Wales is less clear, although religious allusions certainly formed a part of trade union rhetoric. Thomas Halliday was fond of referring to the 'good old Book' in his speeches,[15] while the *Workman's Advocate* lambasted the employers as 'Pharasaical hypocrites' during the lockout of 1875.[16] However, to talk of a Christian-social movement along the same lines as in the Ruhr before the 1890s is erroneous. South Wales lacked such a self-conscious Christian-social movement. Instead, religious references by trade union activists in south Wales reflected a general desire for legitimacy in a religious society, rather than the product of a thought-out ideology. In fact, one historian has suggested that the emphasis on self-help was a hindrance to trade union development.[17] In this interpretation secularisation would have encouraged organisation. But Nonconformist theology did provide legitimacy for the Sliding Scale. Many of the miners' leaders who supported the Scale, such as William Abraham (better known by his bardic name Mabon) and David Morgan, were also actively involved in the religious life of the valleys and had participated in the AAM. Abraham, the staunchest supporter of the Scale and famous for his ability to quell unruly meetings through song, was deeply embedded in the Nonconformist ethos of South Wales.[18] E. Evans has argued that Mabon saw the Scale as holding 'the promise that there should be no more strife in the coalfield, and [he] clearly envisaged an era of peace and justice'.[19] His support for the Scale was, in fact, an attempt to apply his Christian precepts to industrial relations.[20]

The idea of the Scale dovetailed well with Methodist ideas of self-help, as it, in theory at least, ensured that the worker was able to earn as just a wage as possible under the prevailing market conditions. In effect the employers and the employees would share the fortunes of the industry. Mabon shared with the Christian-social movement in the Ruhr a desire to cultivate an understanding between capital and labour. This in no way entailed a complete subordination of the workers' interests. Nonconformity's emphasis on compassion and social responsibility also provided a basis for criticism of employers.[21] The workers were the responsible co-partners of the employers, sharing in the losses and success. When the latter erred, the miners were justified in taking action. Thus Mabon's support for the eight-hour movement was predicated on the idea that restriction of output would raise prices, thereby improving the lot of the men. It was in this vein that he addressed a meeting in Porth in 1894 claiming 'capital and labour, when in intelligent co-operation, should have identical interests, but unfortunately they were not always in harmony, because as a rule capital was not willing to give to labour a fair share of the results of their joint production'.[22]

This preoccupation with a fair share of production was common to both the

Christian-socials in the Ruhr and the Sliding 'Scalites' of south Wales. The 'fair share' formed the basis for accommodation between employers and employees within the discursive practice of both. Each side complemented the other. Neither Mabon nor the Christian-socialists challenged the mineowners' right to the profits of the industry, but merely sought a more just distribution of that wealth. At the same meeting at which he advocated restricted production, Mabon also stressed that the profits should not be so reduced so as to leave the capitalists with no inducement for investment.[23] In both regions, notions of Christian justice and fairness underpinned ideas of cooperation and compromise in industrial relations.

In the 1890s, however, both these cooperative discourses faced challenges from more confrontational interpretations. In the Ruhr there already existed a long-standing conflict between the Christian-social and social-democratic movements, which can be traced back to differences between the *Allgemeiner Deutscher Arbeiterverein* (ADAV) and the Catholic *Knappenvereine* in the 1860s.[24] In 1889 this conflict was submerged by the strike wave that engulfed the Ruhr. The experience of the strike had clearly demonstrated the need for a permanent organisation to represent the miners. Miners' meetings in May resolved that 'the unity of the workers has disappeared and it is now necessary for an organisation to strive for the same'.[25] A delegate conference in Dorstfeld, attended by both Christian-social and social-democratic representatives founded the *Verband zur Wahrung und Förderung der bergmännisches Interessen in Rheinland und Westfalen*, later known as the *Alter Verband*, on 18 August 1889. In an effort to maintain unity, the new association explicitly excluded political and religious matters and by March 1890 the *Alter Verband* had 5,000 members in the Bochum area alone.[26]

This attempt to transpose strike solidarity into organisational unity ultimately proved short-lived. The miners' movement soon split into Christian-social and social-democratic camps. The split between the two had less to do with their concrete goals, than with the specific discourses they applied to industrial relations. For while the Christian-socials continued to seek accommodation between the interests of the owners and the workers, the Social Democrats propagated a more confrontational identity that challenged the authority of management. Instead of being seen as cooperative partners, the miners and the owners were depicted in social-democratic discourse as locked in a struggle for control of the industry.

The ideological gulf between these two discourses was soon demonstrated by a dispute over the political inclination of the Verband's organ, the *Bergarbeiter Zeitung*.[27] The mouthpiece of the association, the newspaper was an especially important means of constructing and disseminating an organisational identity. This made it an in-built point of conflict for the two orientations. The newspaper, dominated by the Social Democrats, promoted a Marxist reading of industrial relations and insisted on the importance of the class struggle (*Klassenkampf*). Improvement for the workers would only come 'through the elimination of the hitherto existing lordship of the mine owners' (*bergwerkliches Herrentum*).[28]

Complaints over the direction of the organ were raised as early as January 1890. A letter in the *Rheinisch-Westfälischer Volksfreund* complained 'that the executive attempts to impose on us, at our own cost, an organ, which in every issue betrays its social democratic character'.[29] At a delegate meeting of the *Alter Verband* in March 1890 the Christian-social deputies from Buer, Hamme and Wattenscheid 'declared themselves decisively against the political viewpoint of the association's organ' and threatened to secede from the *Verband*. A further conference was promised to consider the problem.[30] However, this failed to erase the perception that the *Alter Verband* was dominated by Social Democrats. A meeting of Christian delegates in Bochum demanded that the statutes of the *Verband* be changed so as to exclude Social Democrats from the organisation. The prevailing attitude was 'we want to be a united people in brotherhood, but not under social-democratic leadership ... Our solution must be *Front* against the *Verband*'.[31] On 4 May 1890 a Christian counter association, *Rheinisch-Westfälischer Bergarbeiter Verein Glückauf zur Wahrung und Förderung der bergmännischen Interessen* was founded.

Although *Glückauf* remained comparatively weak, the secession of the Christian-social activists and the ensuing organisational rivalry sharpened the discursive conflict. The Marxist tone of the organ became ever more explicit and attacked the principles upon which its rival was based. Here is an article from May 1890:

> There exists among the working class, and especially among the miners, a mass of people who sing maniacally of the sorrowful nature of their own work and that of their comrades without any reference to economic and political mechanisms. Yet despite the ever more vehement conflict of interest between them and the antagonistic employer class, they dream of a peaceful existence. That might be possible, but then these dreamers are nothing more than the anvil upon which the hammer strikes often and heavily. 'The struggle for existence' will in time be conducted with greater severity and ruthlessness by the industrial circles, and whoever does not devote himself to the same, whoever withholds his energy from his class, perpetrates the class sin of omission, whose consequences will be borne first by himself and his comrades and later by their descendants ... The value of labour will be depressed little by little by capitalist production and will not stop until financial ruin.

This ruin could be countered by organisation, but first every miner had to recognise 'that he belongs to a class, which is ruthlessly exploited by capital'. The only way out of this unenviable position was the development of class consciousness which 'must be supported, must be fostered, otherwise the energetic and successful advocacy of interests will lack the true and only socially relevant mainspring: class camaraderie is a must, a categorical imperative'.[32]

This Marxist reading excluded the possibility of cooperation between the miners and the owners leading to an enduring improvement in the living standards of the former. The mechanism of capitalist economics would inevitably cause conflict. It was far better to accept this and prepare for the clash of interests. Therefore, while the Christian-socials sought a commonality of interests, the *Alter Verband* perceived them as mutually exclusive. Each of the unions had found 'a crystallisation point, which was already ideological, in the one case in the sense of Christian-Catholic confession, in the other in the sense of an intellectually free, socialist world view, which they could apply to the further development of their organisation'.[33] It was these crystallisation points that legitimised action. This was revealed by the January 1893 strike. Meetings of the *Alter Verband* called to discuss its response to a strike in the Saar coalfield were overwhelmingly in favour of a sympathy strike. This was despite warnings from their leader Fritz Bunte that the economic conditions were unfavourable and that they risked punishment for breach of contract. Feelings of class solidarity and empathy were running high. Recommendations regarding financial support were swept aside by cries of 'we have no money' and 'strike'. It was claimed that 'if the comrades in the Saar had the right to strike, then we have the duty to support then'[34] and 'if the local miners do nothing for the Saarbrücker, then one would rightly complain of a lack of solidarity among the miners'.[35] The conviction that there was possibility of harmony between capital and labour was reiterated,[36] and despite Bunte's scepticism a sympathy strike was announced.[37]

The Christian-socialists were extremely critical. *Tremonia* found it 'outrageous that comparatively well paid workers with secure jobs should pick a quarrel during a time of economic decline, without knowing what they actually want, without exactly establishing their demands and without exhausting all other means for their realisation'.[38] It saw the sympathy strike as merely a cynical attempt by the Social Democrats to foment unrest for their own political purposes. The incident merely further revealed the need for an organisation based on a 'Christian basis' to counter the 'agitators'.[39] The Catholic organ was also concerned that the strike would cost the miners the sympathy of the middle class, thereby stalling efforts at social reform.[40] In contrast, the *Rheinisch-Westfälische Arbeiter Zeitung*, a social-democratic organ, claimed that the workers were more than ready to accept reform, but it must be genuine, rather than mere window dressing. Genuine reform, however, could not be expected from the class state.[41] As it was, the strike proved to be a total failure and cost the *Alter Verband* dearly. Only 21,000 of the 145,000 workforce participated. Many activists were arrested and the strikers punished for breach of contract.[42] For the *Bergarbeiterzeitung* this merely revealed the overt class struggle between capital and labour. Protestations of the inevitable victory of the proletariat did little however to ameliorate existing circumstances.[43] By 1894, membership had sunk to 5,144.[44]

In contrast the Christian-social movement made a renewed attempt at organisation the following year. On 28 October 1894 the *Gewerkverein christlicher Bergarbeiter für den Oberbergamtsbezirk Dortmund* was founded under the leadership of August Brust. The son of a miner, Brust was deeply rooted in the Catholic milieu. Following his move to Altenessen in search of work he joined the Catholic *Knappenverein, Bergmannsglück*. He used the association's library to educate himself and also joined the local branch of the *Vinzenzverein*, a Catholic charity organisation.[45] His Catholic socialisation contributed to his vision for the *Gewerkverein*. Ideologically it was committed to maintaining 'peaceful agreement between employers and workers', and a member professed himself 'an opponent of social-democratic principles and aims'.[46] In fact, Brust regarded the organisation as a bulwark against social democracy.[47] The Christian-social ideal of harmonic cooperation between socio-economic groups that underpinned the ideological position of the *Gewerkverein* was expressed through its organ, *Der Bergknappe*. Through the newspaper it also promoted a united stance against the rise of socialism.

> It is the task of all well-disposed workers, but also not least – and this we emphasise – the task of all well-meaning citizens (*Bürger*), including the industrialists, to see to it that socialist tendencies are ever more repressed and that the Christian workforce is united and their unification protected, elevated and allowed for.[48]

The socialists were then the 'other' to be countered by joint action from employers and Christian miners. By 1894 the unity of the 1889 strike had completely vanished. The trade union movement among the Ruhr miners had split into two camps each armed with an effective means to disseminate their conflicting discourses. This served to reinforce the heterogeneity and milieux in the Ruhr.

South Wales seemed in comparison a paragon of industrial peace. HM Inspector of Mines commented in 1890 that a noteworthy feature of the coalfield was 'the harmony and community at large, it is sincerely to be hoped that this good feeling will long continue, for it would be an evil day for South Wales that ushered in a general and protracted struggle in the coal trade'.[49] These might have been famous last words for in the early 1890s cracks began to appear in this harmonious community. But the divisions beginning to emerge in south Wales did not run along religious/social-democratic fault lines. Instead, differences focused on approaches to industrial relations. Ideas regarding collective agreements promoted by the Minerworkers' Federation of Great Britain (MFGB) had penetrated the coalfield, and were able to make headway due to growing dissatisfaction with the Scale since the abandonment of the minimum rate in 1880.[50] In the early 1890s the Monmouthshire and South Wales Miners' Association was formed under the leadership of William Brace. It actively criticised the Scale and advocated affiliation to the MFGB. By 1892 a lodge of the MFGB was established in Mabon's own stronghold, the Rhondda.[51]

The first major challenge to the Scale came in August 1893 with an unexpected hauliers' strike. Characterised as a conflict between 'two rival unions, two union leaders and two diametrically opposed attitudes',[52] the strike starkly revealed the divisions, both occupational and ethnic, that had lain hidden behind the apparent unity of the Scale. Occupational divisions were revealed by the hauliers' exclusion of other types of workers from their meetings. Furthermore, in an effort to protect the legitimacy of the Scale and isolate his opponents, Mabon characterised the strikers as trouble-making English migrants and placed these in juxtaposition to the moderate, diligent, native Welsh miners. At a tumultuous meeting in the Rhondda on 11 August 1893 Mabon was interrupted and insulted by the crowd. In reply he claimed that 'such conduct was something new in a Welsh audience. Welshmen gave fair play (loud cheers). They had never known an Englishman being refused a hearing in Wales but their friend Daronwy had been refused a hearing on the previous day at Tonypandy.'[53] The implication was that an English element with foreign ideas had tainted the normally moderate Welsh miners. Less obliquely, Mabon issued a threat, when he claimed that 'the present strike is not due to Welsh colliers, but a minority of Englishmen at the collieries who have coerced the majority and the owners should consider the position of these afterwards'.[54]

The strike did, however, stimulate debate over the effectiveness of the Scale in protecting the miners' interests. Mabon, David Morgan and other older miners' leaders were forced onto the defensive. At a meeting in Merthyr, Mabon and Morgan presented their case to the miners. The presentation played upon the idea of the uniqueness of the Welsh coalfield. Morgan claimed that the view of the MFGB was 'sit down Morgans, close your eyes and accept what's good for you'. Notions of Welsh nationalism and pride were also pandered to. Morgan believed, for example, that the English leaders were actually jealous of the success of the Welsh trade and, therefore, sought to control it. The main concern was with bread and butter issues and the Welsh leaders 'could not agree with the foundations of the Federation', for the Welsh coalfield was primarily export rather than domestically driven. But underlying these economic differences were two different approaches towards industrial relations. Mabon argued along with 'the members who supported the Sliding Scale, that the principle of measuring wages to the price of coal was a fair and just principle'. The MFGB 'believed wages should control the price of coal ... they differed over the question of principle that was an essential foundation for a common union'. This difference of principle also extended to industrial disputes, for Mabon argued that in Wales they sought to solve quarrels, whereas in England they adhered to the notion of 'paralysing trade' to force the employers to give them what they believed was right. Such an attitude was self-defeating as it was not concerned with the justice of a demand, but was merely a trial of strength and all would suffer the consequences of lost trade.[55]

Mabon perceived Brace's interpretation of industrial relations as having a more confrontational basis. By putting wages before prices the 'Federationists' were impinging on the employers' right to profit. By referring to these ideas as 'foreign' and 'English', Mabon was implicitly depicting them as somehow alien to an imagined Welsh ethos, which was conversely portrayed as conciliatory and cooperative. This dichotomy lay at the centre of his discursive defence of the Scale. Unlike the MFGB, 'the Welsh leaders did not believe in everything for me and the rest for my brothers. They believed in John Stuart Mill's saying that capital and labour should be balanced.'[56]

The MFGB case in South Wales was perhaps most cogently expressed in the pamphlet *The Miners' Evangel.* Written by M. Severn, a Social Democratic Federation (SDF) organiser in Pontypridd and later SWMF activist, the pamphlet was published two years after the hauliers' strike. It is a particularly good example of the mixture of both socialist and Christian thought that underpinned the ideological stance of the MFGB towards industrial issues.

The first chapter of the tract introduced the ideas of socialism through a conversation between father and son. In chronicling the development of socialism, Severn moved effortlessly from Jesus and the disciples as the first socialists through St Simon to Lassalle and Marx.[57] Perhaps more importantly, and in direct contrast to Mabon, Severn explicitly rejected Mill's theory of wages. Instead labour was regarded as the sole creator of wealth.[58] From this basic principle Severn criticised the Sliding Scale as wrong as it 'inculcates that miners' wages must be wholly at the mercy of supply and demand, and places Labour under the heel of Capital'. Instead he argued 'selling prices of coal must be governed by miners' wages'.[59] As in MFGB policy, wages should come first. Dependency between the miner and coalowner was also rejected.[60] In fact, Severn believed that organisation of both the employers and employees would eventually lead to unification of both. However, this union involved the gradual elimination of capital for 'when labour is nationally organised then will labour be competent to compel the instruments of work not to be the monopoly of capitalists, but the common property of society'.[61]

Severn, along with Brace and other Federationists, was, therefore, attacking the key theme within Mabon's discourse. Common interest between owners and miners was rejected in favour of competition over the fruits of the industry. Although this discourse had some similarities with that of the *Alter Verband*, more striking are the differences. For while in the Ruhr religious and Marxist conceptions of industrial relations remained relatively hermetically sealed, in south Wales there was a less doctrinaire blending of the two. This both encouraged and reinforced a more open discursive structure, while in the Ruhr, boundaries between ideological currents were comparatively closed.

## New unionism and new course, 1895–1909

The year 1895 represented something of a caesura for the *Alter Verband*. As a result of its failed strikes its membership had declined precipitously and it was facing increased competition from the *Gewerkverein*. The District President of Arnsberg reported in 1896 that 'while the social democratic association shows a decrease in membership, the Christian *Gewerkverein* has steadily and significantly increased'.[62] The *Alter Verband* was dealt a further blow by the Essen perjury case (*Essener Meineidsprozess*), which resulted in the imprisonment of many of its leaders, such as Johann Meyer and Ludwig Schröder.[63]

From 1895 the *Alter Verband* embarked on a period of reconstruction under the new leadership of Heinrich Möller. This new course also included a modification of the organisation's ideological standpoint. The earlier confrontational tone of the association was diluted and a greater emphasis placed upon working towards practical reform. The most prominent and, from his position as editor of the *Verband* organ, influential proponent of this new discourse was Otto Hue. Through articles and pamphlets Hue continually promoted the idea of neutral trade unions. The most cogent expression of this discourse was his booklet,

*Neutrale oder parteiische Gewerkschaften?*. The central theme was the exclusion of party politics and religious questions from the association. Through strict adherence to purely economic goals, Hue hoped to provide a common basis upon which the trade union movement could unite. Although he repeatedly accused the Christian-socials of fanaticism, Hue freely admitted that earlier the *Bergarbeiterzeitung* had been avowedly socialist. Therefore, he claimed that the historian 'must hold the leadership of the *Verband* newspaper responsible for some of the confusion in the organisation and its decline'.[64] Reconstruction, however, coupled with a deeper study of economics, had placed the *Alter Verband* firmly in the bounds of neutrality.[65] To strengthen his case for cooperation, Hue indicated that the practical demands of the two unions were virtually the same. Both aimed at a decent standard of wages, an eight-hour shift including winding time, workers' safety inspections, workers' committees, the prohibition of female work and changes to the *Knappschaft*.[66] By focusing on the economic interests of the miners Hue attempted to create an identity of the miner as just that, rather than as a social-democratic or Christian variant.

The new emphasis on practical gains and cooperation bore fruit in the late 1890s. In April 1899 the *Bergknappe* reported on the *Alter Verband*'s general meeting in a positive manner: 'In general we must admit that, this time, the tone of discussion was rather quieter, the usual insults directed at our organisation were omitted and every effort made to deal with pure trade unionist issues.'[67] During the *Knappschaft* elections later the same year both unions stood successfully on a joint platform of reform.[68]

Underpinning this example of cooperation was not only a change in attitude within *Alter Verband* discourse, but also a development of a more confrontational stance by the *Gewerkverein*. In the eyes of the *Gewerkverein* leadership its commitment to peaceful relations with employers distinguished it from the *Alter Verband*. The *Essener Volkszeitung*, another Catholic organ, deplored the fact that,

> one now knows that unfortunately in many areas the Christian and the 'red' miners are still being thrown into the same pot. In our opinion this is also the mistake of the Executive of the Mining Association. The Executive can be assured, that in the present case, it deals with, in the Executive of the Gewerkverein, peaceful, level-headed, and moderate men with whom it is not difficult to manage.[69]

The employers, however, were largely unmoved by this attempt to create a dichotomous relationship between the dangerous 'reds' and the practical *Gewerkverein* leadership. From the beginning, the employers regarded the two movements in much the same light and would not allow themselves to be misled 'by the word "Christian", which glosses over Socialism'.[70] This attitude towards the *Gewerkverein* was made plain by the Piesberger strike of 1898 over the issue of Catholic holidays. Just as radical class discourse had contributed to the *Alter Verband*'s support for the Saar miners five years earlier, the *Gewerkverein* now found itself compelled by its religious sentiments to provide financial aid for the Piesberger workers.[71] The uncompromising attitude of the management demonstrated to the *Gewerkverein* leadership the obstacles their conception of industrial cooperation faced. Brust commented:

> The strike and its history demonstrate the audacity and ruthlessness with which certain employers seek to destroy the legally guaranteed rights of the workers. The idealistic property of humanity, Christian feeling and Christian conviction, is sacrificed to the capitalist god, Mammon. On the one hand it is said: 'one should preserve the religion of the people', and 'people of Europe, protect your holiest interests'. On the other hand, the religious sentiment of the worker is destroyed through Sunday and holiday work, for the sake of the ever fuller purse.

Brust also argued that the strike showed the need for the sympathy of the middle class and the need for a stronger organisation.[72]

For Hue the strike was a stage post in the development of the *Gewerkverein* from a 'non-fighting' to a strike association (*Nichtkampfverein zum Streikverein*).[73] Undoubtedly, the *Gewerkverein* was beginning to realise that appeals to Christian sentiments alone were not enough to bring the employers to the table. Heinrich Imbusch, editor of the *Gewerkverein* organ, rejected the idea that the *Gewerkverein* had ever been a 'Nichtkampfverein'. 'It recognised that the employers and the workers have in fact parallel interests, but also that their interests not infrequently stand in opposition ... The strike was for it the last resort.'[74] Religion, however, remained the centrepiece of the *Gewerkverein* identity, as did the desire to establish dialogue. Brust claimed that 'it is an important lesson, which the employers should draw; that it is much better to negotiate with peace-loving workers or their organisations, than to direct them from above'.[75]

The unions were forced into closer cooperation by the 1905 strike.[76] A spontaneous action, it began, like the 1889 strike, at an individual mine, this time the *Bruchstraße* pit, and quickly spread to other mines. Since initial attempts to contain the strike failed, the union leadership eventually established a 'Commission of Seven' to direct the movement and pooled their strike funds.[77] Sympathy was forthcoming from other sections of society. The District President of Düsseldorf reported on supportive meetings of citizens and noted that in Mülheim an evangelical priest even spoke at a social-democratic meeting.[78] This sympathy coupled with the disciplined conduct of the miners meant that employers' calls for military involvement were rejected,[79] and contributed to the introduction of reform. The subsequent mining law of 14 July 1905 abolished *Nullen* and introduced compulsory workers' committees to the mines, although the discussion of wages was expressly excluded from their functions.[80] Hopes for unity between the *Alter Verband* and *Gewerkverein* were maintained by the continued existence of the 'Commission of Seven' and the Prussian miners' delegate conferences of 1905 and 1906.

In south Wales ideas of peaceful cooperation were also put to the test in 1898 by a six-month-long strike. The causes lay in discontent over the level of wages under the Sliding Scale.[81] For the owners the men were the 'aggressors',[82] while the *Colliery Guardian* deplored the 'the obstinacy of the workmen'.[83] In this atmosphere of mutual antagonism Mabon's language of cooperation seemed untenable. Furthermore, the refusal of the strikers to grant Mabon and his colleagues plenary powers to negotiate a settlement suggests that his perspective on industrial relations had lost currency among the workers. In fact, some of Mabon's peers, among them David Morgan, had become convinced of the need for a new form of organisation. Signs of Morgan's conversion came as early as 1894 when he cooperated with Brace in the Society of Colliery Workmen of the South Wales Coalfield.[84] Morgan became more critical of the notion of a co-operative relationship between different classes in society, suggesting that others supported the Scale to the detriment of the miners. Rhetorically he asked the 'landowners, colliery owners and managers, shopkeepers, publicans and 75 per cent of the preachers wished to keep the Sliding Scale for the sake of peace, but had it brought peace to the workmen in the past? (Cries of No, no!)'.[85]

Mabon initially fell back on drawing a distinction between the natives and the 'foreigners'. At Penygraig he claimed 'a new generation had arisen, and strangers had come among them prejudiced against it (the Scale)'.[86] But the experience of defeat convinced Mabon himself of the need for a new organisation. The loss of Mabon's Day, which the employers regarded as a 'valuable point',[87] represented a symbolic defeat for his discourse. He wrote that:

> the settlement is, to my mind, the most unsatisfactory one that has ever been concluded in the district. I don't mean in its intrinsic monetary value, but it fails in the one essential thing for this district – it will not bring the peace necessary for this district to bring it to its full state of prosperity. It is not based on either justice or goodwill. It fails in this important element because the employers insisted upon demanding their 'pound of flesh'.[88]

Let down by his 'partners', Mabon embraced the 'foreign' principles he had previously castigated to become the first president of the SWMF. His erstwhile opponent Brace became vice-president.

Mabon's abandonment of his attempts to 'other' the 'English' element was a large step. Undoubtedly, on one level, Mabon simply sensed the way the wind was blowing and adapted accordingly in an effort to protect his position. More subtly, however, the viability of his initial strategy was limited owing to the extent of in-migration into south Wales. As noted in the preceding chapter, English migrants were comparatively easily assimilated into the mining communities and consequently influenced their development. Unlike the Ruhr, where milieux and trade union discourse were mutually reinforcing, the more cohesive structure of the south Wales mining communities made it difficult to construct a specific identity based on ethnicity or religion. Mabon himself represents this more successful assimilation. Well known for his ability to quell angry crowds through Welsh hymns, he was also just as eloquent in English. Finally, the duration of the strike and the intransigent attitude of the owners emphasised differences between workers and employers. Ultimately, it was the idea of the 'miner' and his relationship to the owners that took precedence over any perceived cultural differences. Whereas trade union discourse in the Ruhr was milieu-fixated, in south Wales it sought to represent the whole community of miners.

Despite this, there seemed in the late 1890s and early 1900s to be a convergence of the various discourses in the two areas. However, while the SWMF successfully united both the supporters of the Sliding Scale and the MFGB, cooperation between the *Alter Verband* and the *Gewerkverein* remained the exception rather than the norm. Periods of joint action were invariably followed by a breakdown in relations and more intense efforts to distinguish one from the other through discourse. This conflict reached a high point during the 1904 *Knappschaft* elections. The *Alter Verband* made its bid for representation of the 'free' miners by terming the *Gewerkverein* a *Zechengewerkverein*, effectively a yellow union. The slogan was 'down with the friends of management and their helpers'. Furthermore the *Alter Verband* argued that it now represented the old Christian-social demands (i.e. reform of the *Knappschaft*).[89] A booklet on the *Knappschaft* elections published six years later recalled the 1904 elections as 'a decisive turning point in the struggle of the miners over their *Knappschaft* rights. The field of battle was cleared; there were no longer three parties on the *Knappschaft*, but only two: the real opposition under the leadership of the [*Alter*] *Verband* and the mines' party under the leadership of Brust.'[90] The *Gewerkverein*, meanwhile, focused on the links between the *Alter Verband* and the SPD, a task made easier by Hue's election to the Reichstag as an SPD deputy for Bochum in 1903.[91] Their counter-slogan was 'down with the Social Democrats across the board'. They claimed the election of *Alter Verband* candidates would merely make the employers more 'obstinate' with regards to reform of the *Knappschaft*.[92] Similarly, while there was agreement on most issues at the first delegate conference of Prussian miners in 1905, at the second the following year political differences between *Alter Verband* and *Gewerkverein* representatives began to resurface.[93]

In 1902 the establishment of a Polish ZZP further fragmented the trade union movement. Both the *Alter Verband* and the *Gewerkverein* had attempted to organise the Polish immigrants, which by the 1900s made up a significant proportion of the mining workforce. However, their attempts were rather lacklustre. The *Gewerkverein*, for example, resisted pressure to produce Polish language literature until 1902. The *Alter Verband*, due to its more internationalist outlook, was less adverse to such methods, but it also exhibited a patronising attitude towards the Polish migrants. In 1899 a strike, accompanied by violence and rioting, occurred in Herne. The main protagonists were young, migrant workers.[94] Commenting on the strike, Otto Hue referred to the migrants as a 'flood-tide of refuse'.[95] The *Alter Verband* organ inconsistently portrayed them as the unwitting defence troops of capitalism, while at the same time claiming them as the brothers of the German workers.[96] Furthermore, the *Gewerkverein* position on the Mining Language Ordinance was that the Poles 'will have to accommodate themselves to existing conditions, whether they want to or not'.[97]

Such attitudes underpinned the Poles' decision to form a separate Polish union. The initial call to organisation came from *Wiarus Polski*. The newspaper played upon the issue of nationalism, requesting Poles not to join any German associations for fear of Germanisation.[98] Soon the union formed part of the Polish subculture in the Ruhr as much as the *Gewerkverein* formed part of the Catholic one.[99] Its statutes excluded religious questions and social-democratic agitation.[100] Its activists claimed that

> the Poles in foreign parts also have the duty to organise. Everyone should affiliate to the union organisation and not let themselves be misled by the old Social-democratic association. You can neither join the Christian Gewerkverein. The Polish working class can not belong to such unions, which cannot tolerate each other and conduct a mutual war.[101]

A former leader, Mankowski, recalled that 'when we left our fatherland we took two great pearls with us, the language and the beliefs of the fathers'. These pearls were endangered by both the socialist and religious trade unions both of which 'would like to draw the Polish worker in – not to protect them, but to take their money and to tear from them the ideas of the fathers. That was the essential inducement to the foundation of our trade union organisation.'[102] Another competing identity had emerged to challenge those of the *Alter Verband* and the *Gewerkverein*, this time predicated on a distinct Polish identity.

John Kulczycki, however, argues that focus upon the role of *Wiarus Polski* in the formation of the ZZP emphasises its nationalist tendencies at the expense of its class character.[103] Certainly, the ZZP soon began to develop a more confrontational identity. Imbusch darkly warned after the 1912 strike, in which the ZZP cooperated with the *Alter Verband*, that 'inflamed nationalism' prevented the ZZP from falling into the arms of the Social Democrats. 'The Polish association and also the Polish Party stand at a turning point. That has again been shown by the strike. Either they will turn back, or they will go under the red flood in the Ruhr.'[104]

The solidarity of the 1905 strike did not last long. Soon after the strike Heinrich Imbusch warned the miners 'against Social Democrats, since they only wanted to spread hate among the masses'. He also focused on the cooperative element in *Gewerkverein* discourse, claiming that, 'had the Social Democrats had the upper hand in the strike, the Government would not have conceded anything and the middle classes (*Bürger*) would not have stood on the workers' side'.[105] The *Gewerkverein* also called a closed meeting to impress upon its members that 'they belong to a separate trade union and should not allow themselves to be influenced by members of other organisations under the cloak of unity'.[106]

The *Alter Verband* also sought to distinguish itself from the *Gewerkverein* by emphasising its oppositional role vis-à-vis the employers. The *Bergarbeiterzeitung* happily reported Emil Kirdorff's statement, 'I do not recognise the Christian trade unions as representatives of the workers. I do not negotiate with them.' However, in response to his claim that the Christian unions were more dangerous than the Social Democrats, the paper retorted that

> obviously Herr Kirdorff does not mean to say the Christian trade union endanger the employers' profit. We have often seen that management and *Gewerkverein*

> candidates go hand in hand against the Verband's candidates during the Knappschaft, mining courts and similar elections ... From this it is apparent which organisation the owners actually consider more dangerous.

This attempt to maintain the integrity of its identity as the sole opponent of the employers was coupled with an explanation of the differing interpretations of industrial relations. The *Alter Verband*, it was claimed, did not hold the employers personally responsible for social problems, as did the *Gewerkverein*. Rather they saw them as the effect of a false system (*verkehrtes System*).[107]

That their doctrinaire ideological viewpoints were the obstacles to cooperation was fully recognised by the unions themselves. The leadership of the *Gewerkverein* claimed that the organisation had, since its founding, struggled against both the employers and the Social Democrats and their trade union movement, 'especially the social-democratic miners' association. This struggle between the two orientations corresponds not so much to actual economic goals, than the difference of ideology which each represent.'[108] For Imbusch amalgamation of the unions was impossible as the *Alter Verband* stood on a materialistic basis, the *Gewerkverein* on a religious one.[109]

Despite Hue's emphasis on neutrality and reform, his conception of industrial relations was still framed in Marxist terms, the very materialism that Imbusch was so critical of. He saw it as a natural consequence of the capitalist system that two antagonistic classes would emerge,[110] and ridiculed the idea of common interest predicated on Christian feeling.[111] In this respect, Hue's discourse failed to transcend that of the earlier editions of the *Bergarbeiterzeitung*. Furthermore, Hue had a blind spot concerning the role that religion played within *Gewerkverein* ideology. He was deeply critical of the capacity of religion to provide solutions to social problems.[112] Religion, however, was the keystone of *Gewerkverein* thinking. Georg Werner, during his attempt to establish a fireman's union (*Steigerverband*), participated in classes held by the Christian union. 'In this way I got to know the religious bonds and reasons, which formed the basis of the ... Christian trade union movement.'[113]

The importance of trade union classes and other organisational factors should not be overlooked. The *Alter Verband* introduced courses on opposition trade unions, in which the *Gewerkverein* was listed alongside the company unions.[114] Although figures are not available for other areas and bearing in mind the economic incentive of strike pay, in Recklinghausen at least it seemed that the *Gewerkverein* did have cause to worry after the 1905 strike. During the strike the local membership of the *Alter Verband* increased from 250 to 1,000, while that of the *Gewerkverein* had stagnated at around 400. Similarly, out of 29 meetings held between 14 January and 4 February, 45 speakers had been Social Democrats compared to 14 Christian-socialists.[115]

Fears of being overwhelmed by the *Alter Verband* spurred the *Gewerkverein* to remind its members of their obligations to its own association. The *Gewerkverein* was placed in the difficult position of avoiding the label of being a yellow union, while at the same time maintaining its independence from the *Alter Verband*.[116]

The specific form of trade union organisation also contributed to the Ruhr unions' inability to overcome their ideological rivalry. The basic unit of organisation, the *Zahlstelle*, was based in the residential districts and members could come from up to 40 different mines.[117] Both the *Gewerkverein* and the *Alter Verband* were also committed to becoming centralised organisations early on.[118] An increasing bureaucratisation paralleled this centralisation. The number of *Alter Verband* officials rose from four in 1894 to 94 by 1914.[119] The individual *Zahlstellen* were given little room for independent action and the need to educate the members in trade union practices was constantly reiterated.[120] The SWMF, on the other hand, was a federal organisation and proposals for centralisation were resisted.[121] This allowed the districts a great degree of freedom and provided a fertile ground for the development of new ideas. Furthermore, the pit lodge system based organisation around individual pits, placing activists in a better position to exert pressure on non-unionists. One means was the show-card day. This method clearly identified non-unionists to their organised colleagues. Non-unionists were even visited at home.[122] Within both the workplace and the community, pressure could be placed upon individuals to join the union. The siting of the *Zahlstelle* in the residential districts, on the other hand, strengthened ties to the individual milieux, be they Catholic, Polish or Social Democratic. They became strands in the dense web that made up those milieux. In south Wales the concentration of organisation on the workplace helped strengthen the SWMF's claim to be representing the miners as a whole, rather than a particular religious or ethnic group. This discourse helped the SWMF achieve its high organisational density soon after its formation (see Table 2.1). Once achieved, this high density appeared to further legitimise the leaders' claims as they could confidently assert that they represented the majority of the workforce. It was therefore possible to forge a more direct and organic link between workplace and organisational solidarity.

The position of checkweigher in south Wales also provided a natural point of leadership for a workforce. Directly elected by the workforce and paid for from their wages, the checkweigher was at least in theory largely independent from managerial control. The importance of this office was fully recognised by employers.[123] The manager of Locket's Merthyr Collieries reported approvingly of one checkweigher, 'he is not a demagogue or busybody' only to later deplore the election of T. Isaac Jones because he was 'an ex-student of Ruskin College, Oxford – a great opponent of Mabon; a gentleman who, I fear, will cause considerable uneasiness amongst the men and annoyance to the company and its officers'.[124] The mines of the Ruhr lacked an equivalent position. Admittedly, the workers' committees and elected safetymen (*Grubenkontolleure*), introduced in 1905 and 1909 respectively, could have provided an organisational focus.[125] Initially, however, the *Alter Verband* boycotted the first committee elections, complaining that they had no power to discuss wages. Later it participated on the committees, although doubts about their effectiveness as a means of representing the workers' interests remained. Only a strong union organisation could make them truly effective.[126] Furthermore, the office of safetyman was criticised

*Table 2.1* Membership of the main trade unions in the Ruhr, 1898–1913

| | Total | *Alter Verband* | | *Gewerkverein* | | ZZP | |
|---|---|---|---|---|---|---|---|
| Year | employed | No. | % | No. | % | No. | % |
| 1898 | 191,847 | 17,974 | 9.4 | | | | |
| 1900 | 226,902 | 12,945 | 5.7 | | | | |
| 1910 | 345,136 | 76,418 | 22.7 | 33,000* | 9.6 | 26,309 | 7.6 |
| 1912 | 393,879 | 69,648 | 17.7 | 40,000* | 10.2 | 30,354 | 7.7 |
| 1913 | 394,569 | 62,487 | 15.8 | | | 28,936 | 7.3 |

*Approximate figures.
*Source:* Figures compiled from Hickey, *Workers in Imperial Germany*, p. 234; Kulczycki, *Foreign Worker*, p. 218; Brüggemeier, *Leben vor Ort*, p. 283.

*Table 2.2* Membership of the SWMF, 1898–1913

| | | South Wales Miners' Federation | |
|---|---|---|---|
| Year | Total employed | No. | % |
| 1898 | 128,313 | 60,000 | 47 |
| 1900 | 147,652 | 127,894 | 87 |
| 1910 | 213,252 | 137,553 | 65 |
| 1912 | 225,535 | 114,208 | 51 |
| 1913 | 233,134 | 153,813 | 66 |

*Source:* Williams, *Capitalism, Community and Conflict*, pp. 88–9.

as mere palliative (*weisse Salbe*), subject to pressure from the management to find 'everything in order'.[127] The *Gewerkverein*, on the other hand, was more positive and regarded safetymen as a genuine 'step forward in the improvement of miners' safety'.[128] The *Gewerkverein* was aware of the limitations of the committees and 'mistrust, if not bitterness' with which the employers regarded them. But the union hoped they would provide a basis for better relations. They believed that through the committees

> the workers and the employers will again get to know and understand each other. Then the employers will have to realise that the workers are better and more moderate than they have until now imagined and the workers will come to see that the owners are not so consistently unreasonable (*grundschlecht*) as they often believe ... The result is a mitigation of class antagonism.

It was precisely because of this that the Social Democrats were said to oppose the committees.[129] So while for the *Alter Verband* the committees and safetymen were mere decoration, for the *Gewerkverein* they were the potential means of establishing a cooperative relationship between the workers and the employers.

Ultimately, rather than providing foci for mine-based organisation, the committee and safetymen elections became another level of competition between the trade unions. The *Alter Verband* claimed that only the election of its members could prevent the safetymen from becoming *weisse Salbe*.[130] It even produced a booklet instructing the safetymen of their duties. It was hoped that through inspections they could influence wages.[131] The *Bergämter* also received complaints of preferential treatment meted out by safetymen of one union to fellow members.[132] Interest in this position soon began to wane as it carried the double risk of both ostracism by their workmates and pressure from management.[133] Furthermore, the two unions' attitude towards the works' councils and safetymen were directly linked to their discursive interpretations of industrial relations. While the *Alter Verband* eventually came to see these offices as a chance to exert some form of influence on the management of the mines, the *Gewerkverein* regarded them as a step towards genuine cooperation between miners and owners. While the first union hoped to use them in a combative manner, the latter believed they could be used to encourage dialogue. Unsurprisingly, during the works committee elections each union took the opportunity to campaign for its own members and denigrate its opponents.[134]

Organisational differences alone, however, cannot account for the antagonism between the two trade union movements. Ultimately, the discursive identity created by the SWMF proved more flexible than that constructed by the unions in the Ruhr. This flexibility was evident from its very inception. By giving the presidency and vice-presidency to 'Mabon, as the most successful of the older regime of leadership, and … to Mr Brace, the leader of the Federation campaign',[135] the miners were not so much breaking with the past as signifying the coexistence of different identities within the organisation. Mabon himself continued to describe himself as a 'conciliatory' spirit,[136] but his religious metaphors were now employed in the service of the SWMF. Fears that the Revival which swept south Wales in 1904 would damage the SWMF were refuted by Mabon, who claimed that:

> the men that are as a rule prepared to stand out for labour rights are the men in our churches who stand out for religious rights … godly men – good men – are convinced that although God in a sense provides for all, that ungodly men take more than their fair share of what He has provided, and we believe that God Himself really blesses those who provide for themselves … we [must] have good and godly men as mainstays to our organisation.[137]

Through his study of strike statistics, Deian Hopkin has identified an increasing militancy among the south Wales miners from 1898,[138] and this was reflected by the development of a harder edge to Mabon's language. Cooperation might have remained his preferred policy, but the intransigent attitude of the owners during 1898 had also convinced him of the need for force. During agitation for the Eight Hours Bill Mabon claimed that 'we (the miners) can appeal, if necessary, to our Father, Who is in Heaven, to defend the right, and we ask you, the public, to defend His children while they are seeking this measure of right and justice'. Meanwhile, Brace struck a more prosaic, yet nonetheless confrontational tone.

> Now that we have carried this agitation to this point, we are not going to allow the enormous power of vested interest to drive us back. If the coalowners of Britain, backed by interested consumers who do not recognise that, after all, underlying this

great reform there is a great principle of high humanity – if they allow us to be defeated on this occasion, so far as I am concerned – and I think I am speaking on behalf of my colleagues – we shall dedicate no more portion of our time to get it from Parliament, but we shall be parties in using the power of our great combination (loud cheers) to force the recognition of this much too long delayed reform.[139]

Both religiosity and humanism acted as a justification for the Eight Hours Bill, even to the point of a strike, although the union did not resort to industrial action. This double legitimacy provided it with the widest possible appeal. In the Ruhr, on the other hand, the unions were based on ideological 'crystallisation points'. Each point, be it Christian-socialism, social democracy or ethnicity, was linked to specific milieux or sub-cultures. Since membership of these milieux was generally mutually exclusive it was difficult for trade union activists to combine different discourses. The adoption of a certain discourse mobilised a specific group of miners, but to blend discourses risked alienating their original members. The unions also attempted to impose harmony in their organisations by restricting pluralistic debate. Therefore, in terms of discourse, it became a case of one or the other. Meanwhile, the lack of milieux in south Wales meant that activists could unite various discourses without fear of potentially alienating a significant section of the workforce. The leadership's acceptance of pluralistic debate allowed the creation of a more inclusive, elastic discourse that sought to represent the miners as a whole.

## Syndicalists and anarchists, 1909–1914

Towards the end of the first decade of the twentieth century syndicalist thought began to make an impact in both the Ruhr and south Wales. Contemporaries regarded strikes around 1910 and 1911 in both regions as influenced, if not caused, by syndicalist ideas. Syndicalist influences and currents have been identified within both the British and German labour movement from the 1890s.[140] The precise role that syndicalists and their ideas played in industrial relations is, however, notoriously difficult to pin down. In Germany the movement had a more concrete organisational base in the *Freie Vereinigung deutscher Gewerkschaften* (Free Association of German Trade Unions, FVdG), which explicitly adopted the general strike doctrine in 1904,[141] at about the same time the social-democratic movement was debating the efficacy of the mass strike.[142] Syndicalism's role in Britain is comparatively more nebulous and undefined. Although syndicalist bodies like the Industrial Syndicalist Educational League were formed, they tended to exercise influence through the existing trade union structures.[143] That they were able to do so again illustrates the comparative elasticity of trade union discourse in Britain.

These national patterns were replicated in both south Wales and the Ruhr. The Cambrian Combine dispute formed the background to the emergence of the Unofficial Reform Committee (URC), a loose body of individuals representing different political and industrial ideologies.[144] Significantly the URC was not an organisational alternative to the SWMF, but rather a platform for reform. At its earliest meetings older members of the SWMF such as Vernon Hartshorn and James Winstone were present.[145] Furthermore, William F. Hay, one of the staunchest syndicalists, advised men at several Rhondda collieries who had seceded from the SWMF due to discontent over the 1910 Conciliation Agreement to rejoin the organisation, while at the same time acknowledging the necessity for reform.[146] On the other hand, the *Freie Vereinigung der Bergarbeiter* (FVB), founded in 1910, was an independent trade union with its own statutes. It was affiliated to the FVdG and subscribed to its organ, *Die Einigkeit*. As such it represented an alternative to the *Alter Verband* and *Gewerkverein*, albeit a numerically unimportant one.[147]

In south Wales the issue at stake in 1911 was the minimum wage rate. Marxist, syndicalist and industrial unionist thought influenced many graduates of Ruskin College, and later the Central Labour College (CLC). Upon their return to the valleys they sought to spread these ideas through evening classes on industrial history and economics.[148] Charlie Gibbons, a protégé of Noah Ablett, wrote of his time at the CLC, 'I was taught ... that the interests of the workers are absolutely opposed to the interests of the employers ... [and] a fight must be waged until the employers are eventually wiped out'.[149] While the older MFGB supporters had placed the workers and owners into competition, they did not necessarily foresee the actual elimination of the employers. The URC went a step further in challenging not only the employers' share of profit, but also their control of the workplace and of the industry as a whole.[150]

In the Ruhr syndicalist and anarchist traditions more decisively influenced the left-radical (*linksradikal*) current. Marxist discourse was still monopolised by the *Alter Verband* and the SPD, although there were affinities between left-wing supporters of the mass strike and the anarchosyndicalists. Unlike the URC, the FVdG rejected politics, although not individual membership of parties, in favour of the economic struggle conducted through general strikes, direct action, sabotage and passive resistance.[151] Also, while the Welsh syndicalists and industrial unionists sought a centralisation of the SWMF to make it more effective, the anarchosyndicalists of the Ruhr actually advocated a move away from central to local organisations.

Despite these organisational and ideological differences the critique of the existing trade union movement shared certain affinities. Chief among these was the lack of fighting spirit. Hay and Ablett stressed the divisive effects of differing minimum rates upon the workers' movement. They stressed that any minimum rate was the common goal of all grades of workers in the mines. This underlying theme of unity was then bound up with recurrent metaphors of struggle.

> Remember, in this work, all petty pride and prejudice has to be dropped. It is a case of each for all, and all for each. The collier or hewer, the timberman, putter, labourer, etc. have all to *fight* for it ... Let us go into battle, then, united at last in one common demand, with one unifying demand, with all the solidarity which that common cause can give. Or demand one, our members as one, our fight as good as won, since we are united.[152]

*The Miners' Next Step* was a blueprint to transform the SWMF into a more effective fighting organisation. That the 'old policy of identity of interest between employers and ourselves be abolished, and a policy of open hostility installed',[153] was predicated on the recognition of 'the war of interest between workers and employers'.[154] Similarly, the FVB claimed that the true policy of the unions should be 'inexorable struggle'. The policy of the central trade unions was criticised as being too conciliatory and cautious. In contrast to the FVB emphasis on conflict, Sachse, the then leader of the *Alter Verband*, was quoted as seeing the purpose of the unions as avoiding economic trouble.[155] The FVB complained that every time the decisive moment for action drew near a central executive appeared to beat it down.[156] In both cases the notion of the class war was more explicitly drawn and the miners placed in the vanguard of that conflict, seeking the destruction of the existing economic system and the establishment of workers' control of the mining industry.

Linked to this more confrontational approach was a critique of the inertia of the existing trade union executives. The FVB focused on 'mismanagement, despotism, autocracy and bureaucracy in the present day central associations'. It argued that the development of the trade union movement showed that it could never lead to the freedom of the workers, but 'leads them from the servitude of capital into the servitude of the central association and [SPD] party popes'.[157] *The Miners' Next Step*, while admitting that leadership had its good points, also attacked the tendency for power to be removed from the rank and file workforce and concentrated in the hands of executive members. Furthermore, leaders hindered the emergence of solidarity based on interests common to all,[158] a prerequisite for the success of the trade unions. William Henry Mainwaring, a co-founder of the URC, launched a scathing attack on trade union policy through the recently created working-class organ *The Rhondda Socialist*.

> We thus see our desired end in an Organisation of the Working Class, which will be strong to fight. So strong, that the opposing Capitalists will be driven to see the futility of fighting. Let it be remembered that the Capitalist fights today, not because he loves fighting; far from it, but fight he must if he wishes to maintain his profits. Hence his desire for Conciliation, because he always hopes to show the Labour Leader – that epitome of ineptitude – that he cannot afford to pay a wage on which a man can live. One could never frame a more scathing indictment of Conciliation than that of a Brace, whining for a peep at the balance-sheets of the employers, in order to salve his conscience for being a party to accepting starvation wages on behalf of those whose interests he is so well paid to defend. No! An Organisation which exists to continually accommodate its members' stomachs to its opponents' pass-book can never serve as the medium for Labour's final emancipation! *Yet it is the property of its members*! As such they should seize it! Renovate it! Transform it! Shape it into a weapon to wrest control of Industry from the Capitalist, later to afford a medium for administering the Industry by the Workers themselves. In a word, to make of it an Industrial Union, with the definite objective of establishing *Industrial Democracy*![159]

Although both syndicalist movements remained small, they did have an effect upon the trade unions. Admittedly, the URC plans to reform the SWMF were rejected by a ballot vote.[160] However, the election of three socialist (but not syndicalist) leaders to the executive and the replacement of Mabon by the coalfield firebrand C. B. Stanton on the executive committee of the International Miners' Federation indicated that the miners were certainly willing to countenance a more confrontational policy than hitherto pursued.[161] Mabon himself believed that his policy of conciliation had been 'rejected unmistakably'.[162] Furthermore, the issue of the minimum wage was taken up by the MFGB. Although the Minimum Wage Act hastily passed by the government left many unsatisfied, in comparison to the action of the same year in the Ruhr, the 1912 national coal strike can be regarded as at least a partial success.[163] The strike also revealed the potential pressure that could be brought to bear on the government by united action by the miners, something that was to be of significance during the Great War. Although debate on policy could degenerate into personal recriminations, it is significant in terms of the comparison that ideological differences did not split the SWMF. Vernon Hartshorn argued that it was natural that different policies should emerge. The SWMF should show tolerance and only very compelling reasons could justify an attempt to remove an official from his post.[164] Therefore, the more radical discourse of the URC filtered into that of the SWMF. Rather than standing in direct opposition, this discourse worked from within to modify existing discursive practices, making them more combative, but not replacing them in entirety.

In the Ruhr, the *Alter Verband* was fiercely critical of the anarchosyndicalists claiming that a return to local organisation would be playing into the employers' hands. Only highly organised miners could counter highly organised employers. 'Whoever wants to help the workers must advise the organisation which alone can line up against the employers; that is the large, stalwart, well disciplined central association.'[165] Both movements saw themselves as engaged in a class struggle, but the manner in which it was to be conducted and the role the individual miner was to play was essentially different. While the FVB saw the miner as the potential perpetrator of direct action, the *Alter Verband* discourse idealised him as the disciplined foot-solider of the organisation.

Despite their small numbers, the anarchosyndicalists contributed to pushing the *Alter Verband* into the wage movement that underpinned the 1912 *Dreibundestreik* (Three Association Strike).[166] Unlike the 1889 and 1905 strikes this was a planned action. However, it was fatally crippled by the refusal of the *Gewerkverein* to participate. Out of three areas, Essen, Bochum and Dortmund, only in the latter did a majority of the workers strike.

Relations between the strikers and those who continued to work were strained,[167] further souring the trade union discourses. Accusations of strike breaking were countered by claims it was a duty not to strike.[168] The *Gewerkverein* action was predicated on their more conciliatory identity. It pointed out that the owners were already considering a wage increase and there was, therefore, no need to provoke a dispute. Instead it was better to wait for the

*Table 2.3* Number of strikers and total mining workforce in Bochum, Dortmund and Essen, 19 February 1912

| | Total workforce | No. of strikers |
|---|---|---|
| Bochum | 24,489 | 7,513 |
| Dortmund | 69,413 | 42,401 |
| Essen | 46,170 | 12,633 |

*Source: Köngliche Bergwerksdirektion* to Minister for Trade and Industry, 19 March 1912, STAM OBA 1857.

employers' response and use this as a basis for cooperation. Provocation could only serve to alienate the public.[169] This peaceful stance was contrasted with the conduct of the *Alter Verband*, which demonstrated its desire for a struggle under any circumstances.[170] The strike was also portrayed as essentially politically motivated and designed to destroy the *Gewerkverein*.[171] The common interest of the German miners and owners regarding the health of the industry also meant that the British strike offered the opportunity to improve Germany's competitive edge in the world markets.[172]

The *Alter Verband*, on the other hand, believed that the time had never been so favourable for a strike. They argued that the owners had enjoyed increased profits from rising prices for some time, but that this had not filtered down to the workers. Through their 'ruthless profit policy' the owners had upset both consumers and workers. The union argued 'the workers would be fools' if they believed employers' assurances.[173] Meanwhile Imbusch's notion that the British strike should be exploited in the interests of German economic interests was fiercely condemned. This proved that 'these "workers' representatives"' stood for the interest of German capital and wanted to strike at the interests of the English comrades' and that 'the principles of the Christian Gewerkverein during the strike movement were completely yellow sentiments, completely yellow principles'.[174]

In the case of both unions their discursive practices acted as the basis of legitimacy for action. Since the *Gewerkverein* regarded the employers' and workers' interests as generally the same they were content with a statement that a wage increase would be forthcoming if the economic conditions allowed. This notion also lay behind Imbusch's attitude towards the British strike. On the other hand, the *Alter Verband* view that the two sides were necessarily in competition meant that such assurances were seen as merely window dressing, designed to distract the miners from the huge profits the owners had already made. Finally, while supporting the British miners was not a primary motive behind the Ruhr strike, inter-linked notions of class identity and the international struggle against capitalism hardened the *Alter Verband* attitude towards the employers. Moreover, this more combative attitude provided a means of differentiating the *Alter Verband* from the *Gewerkverein*.

The influence of syndicalist ideas upon the trade union movements in both regions differed substantially. Both British and German syndicalist critiques focused on similar aspects of trade unionism, such as bureaucratisation. Both also sought to propagate a more confrontational identity to mobilise the miners. However, the former seems to have been more integrated into the existing movement. Discursive differences did not lead to organisational separation. The URC members did not advocate a new organisation, but reform of the existing association. As such they exerted pressure from within. The anarchosyndicalists of the Ruhr, on the other hand, formed their own organisations and pressurised the *Alter Verband* from without. Undoubtedly, this was partly a consequence of the relative strength of the unions at the time. However, the Welsh syndicalists, like the SWMF generally, were also firmly rooted in the comparatively broader and more cohesive culture of the communities they inhabited. On the other hand, in the Ruhr, trade union discourse was closely connected to a particular milieux or sub-culture. The trade unions sought to mobilise these milieux through their discourse, while at the same time making clear distinctions between themselves and their rivals. As a consequence of this desire for clear self-identification, trade union discourse tended to be more inflexible in the Ruhr than south Wales.

## Summary

This chapter has tried to show how the discursive identities created by the trade unionists in south Wales and the Ruhr differed. It is clear that the three discourses highlighted here (Christian-socialism, socialism or social democracy, and syndicalism) were used in different ways. In south Wales the SWMF was able to combine these three strands to create a more inclusive identity. In the Ruhr, on the other hand, each discourse was the preserve of a different union. And each union sought to distinguish its discourse by placing it in juxtaposition with the others. The result was more exclusive identities that were defined by what they were not as much as by what they were.

In part this was due to the nature of the civil societies in which the trade unions operated. As noted in the preceding chapter, there was a ‘pillarisation’ of civil society in the Ruhr into several milieux. The trade unions were an integral part of this segmented civil society in which group membership was mutually exclusive. The *Gewerkverein*, with its provision that members openly accept Christianity and reject social democracy, represented a particularly extreme example of how multiple and overlapping memberships were regarded with hostility. Such an assertion naturally limited the number of miners that the *Gewerkverein* could hope to organise. It also meant that the *Gewerkverein* was unable to utilise aspects of social-democratic discourse to forge its own identity. Indeed, the *Gewerkverein* was careful to distinguish its Christian-socialist ethos from social democracy. It therefore contrasted its desire for peaceful cooperation between workers and employers with the *Alter Verband*’s idea that industrial relations essentially represented a conflict of interests. Similarly, Hue’s conviction that religion was, at best, irrelevant to industrial relations meant that he could not create an identity that encompassed Christian-socialism. He could not come to terms with the extent to which religion was the foundation stone of the *Gewerkverein*. In any case, to attempt to use Christian thought risked alienating the committed Social Democrats that formed the core of its membership. In order to legitimise its claim to be the only true miners' union the *Alter Verband* repeatedly characterised the *Gewerkverein* as a 'yellow' union. In their different ways the leaders of both organisations consistently sought through their rhetoric to differentiate themselves from their rivals, while the centralised structure facilitated the imposition of an accepted viewpoint. Thus, although the unions claimed to be speaking for the miners as a whole, each was in effect presenting an identity that upheld a certain constituency, a 'particularist' ideal of the miner; the one Christian, the other social-democratic. Occasionally, a truce would be called, as during some of the *Knappschaft* elections, or cooperation would be forced on the unions by unofficial industrial action, as in 1905. Significantly, however, these incidences encouraged joint action rather than amalgamation. In fact, the mere hint that one or other of the unions was threatened with losing its distinctive identity led to a renewal of their rivalry. It was such fears that galvanised Imbusch to pen his pamphlet rejecting amalgamation in 1906.

South Wales, on the other hand, suffered comparatively little in terms of religious or ethnic tensions. The mining communities were not divided into various milieux, consolidated by state pressure and hostile to membership of other groups. The result was a more fluid civil society. Admittedly, Mabon had attempted to utilise tensions between the native Welsh miners and the immigrant English in the early 1890s to safeguard the Sliding Scale. But this attempt lacked the same resonance which Christian-socialism and social democracy had with subcultures in the Ruhr, while the six-month strike revealed the weaknesses inherent in such a course. Instead, pluralist discourse was accepted, if not always welcomed, within the miners' movement. For Hartshorn it was only natural that policy differences should emerge within a large organisation. This toleration of pluralism allowed the accommodation of different ideologies and enabled the SWMF to dominate miners' organisation in south Wales. Religiously inclined miners, socialist miners and committed syndicalists could all find their own niche within the SWMF's discourse. Furthermore, the federal and pit-based organisation of the trade union helped it to portray itself as the representative of all miners, irrespective of religious beliefs or ethnic origin, and allowed space for the continued existence of local differences within an overarching framework. The SWMF amalgamated various elements all committed to the shared goal of improving the miners' lot. Tensions between the different ideologies certainly existed and were played out in the press, often becoming personal in nature. But these did not lead to a split, for, unlike the Ruhr, there was no fundamental disagreement in south Wales as to who was represented. The activists, whatever their own ideological position, sought to represent the miners in their entirety. What was at question was the means to improvement, rather than who or what the miners should be.

The above examination of trade unionism has stressed the important role of trade union leaders and activists in constructing and maintaining trade union identities. These identities were not created in a vacuum, but were in a dynamic relation with structural developments. To put it simply, the presence of large numbers of Polish workers in the Ruhr created the potential for the formation of a union based on a Polish ethnicity, but it took the activists and their language of Polish identity to mobilise those workers. Similarly, the *Gewerkverein*, *Alter Verband* and SWMF all provided a framework through which the miners could view events such as strikes, lockouts and wage disputes. This is not to deny the element of coercion that the trade unions could exert on non-members to join. The SWMF by virtue of its large membership was especially well placed to pressure miners to join. However, even the SWMF required a critical mass of membership before it could begin leaning on the unorganised. The doctrinal differences between the *Gewerkverein* and *Alter Verband* meant they were never able to achieve such hegemony over the miners, thereby reinforcing their weakness.

Some of the factors that contributed to the development of trade unionism in both coalfields were undoubtedly specific to those regions. Again the lack of sectarian tension in south Wales can be compared with Scotland. Similarly, questions can be raised over the importance of the Free Trade Unions in the Catholic southern German states. However, the emphasis laid on activists' roles and the parameters set by civil society can be applied more broadly. In an interesting aside the Salford working-class autodidact, Robert Roberts, recalled in his memoirs that he met two committed trade unionists during his apprenticeship. One, a Christian-socialist, argued that profit was a sin, while the second, a committed Marxist, taught him the theory of surplus value. Significantly, although their justifications for unionisation were very different, both were in the same union.[175] Given the segmented nature of civil society in Wilhelmine Germany it is probable such activists would have been in different unions.

In summary, the trade union discourses that did emerge were not simple reflections of milieux. Over time the identities of all the trade unions underwent modification. The *Gewerkverein* and SWMF became more aggressive, while the *Alter Verband* played down its anti-religious aspects. In both south Wales and the Ruhr, discursive identities were as much the product of individuals as they were of social structures. Figures such as Brust, Mabon, Hue, Imbusch, Brace and Hartshorn were crucial to forging these identities. Ultimately, the trade unions in the Ruhr created identities that were comparatively doctrinaire and as such contributed vitally to the shaping and maintenance of the various milieux between 1890 and 1914. The miner was not just a miner. He was a Social Democrat, or a Christian or a Polish miner. It was difficult to move beyond these identities to forge a new one. The different approaches to industrial relations were tightly bound up with who the miner was perceived to be. Was he the religious miner cooperating with the owners in the spirit of Christian brotherhood or the class warrior preparing for the moment when the capitalist system ultimately collapsed? *Knappschaft*, workers' committee and safetymen elections all offered a stage upon which this rivalry could be played out. Only the crisis of war was to push the *Alter Verband* and *Gewerkverein* together.

## Notes

1 H. Marchwitza, *Meine Jugend* (Berlin: Tribüne, 1976), p. 241.
2 Paynter, 'SWMF', p. 68.
3 *Ibid.*, pp. 40–71.
4 See Tenfelde, *Sozialgeschichte der Bergarbeiterschaft*, pp. 470–86 and pp. 514–21 and Imbusch, *Arbeitsverhältnis und Arbeiterorganisationen.*
5 J. H. Morris and L. J. Williams, 'The south Wales sliding scale, 1876–79: An experiment in industrial relations', in W. E. Minchinton (ed.), *South Wales*, pp. 218–31.
6 E. W. Evans, *The Miners of South Wales* (Cardiff: University of Wales Press, 1961), pp. 133–8.
7 Lewis, *Rhondda Valleys*, p. 167.
8 See Evans, *German Center Party*, pp. 178–80 and Imbusch, *Arbeitsverhältnis und Arbeiterorganisationen*, pp. 190–3
9 *Jahresbericht, Verein der Belegschaft der Zeche Ver. Hannibal*, 14 April 1855, cited in Kroker, *Solidarität*, p. 63.
10 G. Adelman, *Die soziale Betriebsverfassung des Ruhrbergbaus vom Anfang des 19. Jahrhunderts bis zum Ersten Weltkrieg unter besonderer Berücksichtigung des Industrie- und Handelskammerbezirks Essen* (Bonn: Ludwing Röhrscheid, 1962), pp. 100–8.
11 *T*, 29 July 1889.
12 L. Lensing, *Der große Bergarbeiter-Streik des Jahres 1889 im Rheinisch-Westfälischen Kohlenrevier. Ein Wort zur Abwehr* (Dortmund, 1889), p. 82.
13 H. G. Kirchhoff, *Die Staatliche Sozialpolitik im Ruhrbergbau 1871–1914* (Cologne: Westdeutscher Verlag, 1957), pp. 91–4.
14 *T*, 11 December 1889.
15 *WM*, 1 June 1871.
16 *The Workman's Advocate* 15 January 1875.
17 K. Burgess, *The Origins of British Industrial Relations: The Nineteenth Century Experience* (London: Croom Helm, 1975), pp. 170–6. Robert Pope also claims that the Welsh chapels' focus on individual salvation restrained their involvement in social questions. See R. Pope, *Building Jerusalem: Nonconformity, Labour and the Social Question in Wales, 1906–1939* (Cardiff: University of Wales Press, 1998), pp. 241–9.
18 See E. W. Evans, *Mabon: William Abraham 1842–1922: A Study in Trade Union Leadership* (Cardiff: University of Wales Press, 1959); E. W. Evans, 'Mabon and trade unionism in the south Wales coalfield', in Hughes (ed.), *Men*, pp. 51–9; and D. A. Pretty, 'David Morgan (Dai O'r Nant), miners' agent. A portrait of leadership in the South Wales Coalfield', *WHR*, 20:3 (2000), 495–531. Mabon was an enthusiastic participant in the Welsh National Eisteddfod, hence his bardic name. He will be referred to as such from here on.
19 Evans, *Mabon*, p. 15.
20 *Ibid.*, pp. 100–1.
21 See A. J. Reid, 'Old unionism reconsidered: The radicalism of Robert Knight, 1870–1900', in E. F. Biagini and A. J. Reid (eds), *Currents of Radicalism: Popular Radicalism, Organised Labour and Party Politics in Britain, 1850–1914* (Cambridge: Cambridge University Press, 1991), pp. 214–45.
22 *SWDN*, 1 October 1894.
23 *Ibid.*
24 K. Tenfelde, 'Bergmännisches Vereinswesen im Ruhrgebiet während der

Industrialisierung', in Reulecke and Weber (eds), *Fabrik, Familie, Feierabend*, pp. 323–4.

25 Cited in Imbusch, *Arbeitsverhältnis und Arbeiterorganisationen*, p. 307.

26 Hickey, *Workers*, p. 228.

27 The *Bergarbeiterzeitung* experienced several name changes during its lifetime. Here it will be referred to as the *Bergarbeiterzeitung* throughout to avoid confusion.

28 *BZ*, 15 February 1890.

29 *Rheinisch-Westfälischen Volkszeitung* 30 January 1890, cited in Imbusch, *Arbeitsverhältnis und Arbeiterorganisationen*, p. 320.

30 *Bottroper Volkszeitung* 11 March 1890.

31 *Ibid.*, 13 March 1890.

32 *BZ*, 17 May 1890.

33 W. Neumann, *Die Gewerkschaften im Ruhrgebiet. Voraussetzung, Entwicklung und Wirksamkeit* (Cologne: Bund-Verlag, 1951), p. 56.

34 *Dortmunder Zeitung* 9 January 1893.

35 *Ibid.*, 8 January 1893.

36 *RWAZ*, 11 January 1893.

37 *BZ*, 7 Jan 1893.

38 *T*, 7 January 1890.

39 *Ibid.*, 9 January 1890.

40 *Ibid.*, 8 January 1890.

41 *RWAZ*, 18 January 1890.

42 Brüggemeier, *Leben vor Ort*, p. 196.

43 *BZ*, 18 February 1890.

44 Neumann, *Gewerkschaften im Ruhrgebiet*, p. 78.

45 See C. Hiepel, *Arbeiterkatholizmus an der Ruhr. August Brust und der Gewerkverein christlicher Bergarbeiter* (Stuttgart: Kohlhammer, 1999), pp. 34–9.

46 Statutes reproduced in Imbusch, *Arbeitsverhältnis und Arbeiterorganisationen*, p. 716. See also H. J. Wallraff, 'Die Belastung einer Gewerkschaft durch ideologischer Differenzen – Spannungen innerhalb der christlischen Gewerkschaftsbewegung in den Jahren 1900–1914', in H. O. Vetter (ed.), *Vom Sozialistengesetz zur Mitbestimmung. Zum 100. Geburtstag von Hans Böckler* (Cologne: Bund Verlag, 1975), pp. 135–52.

47 H. Imbusch, *August Brust. Ein Lebensbild des Gründers der ersten christlichen Gewerkschaft* (Berlin: Christlicher Gewerkschafts-Verlag, 1924), p. 8.

48 *Bk*, 23 November 1895.

49 HM Inspector of Mines South Wales District No 13, p. 1, GRO D/D NCB 67/2/6.

50 Morris and Williams, 'Sliding scale', p. 230

51 *WM*, 9 November 1892.

52 M. Keen and R. Keen, 'The coal war in south Wales, 1893', *Glamorgan Historian*, 10 (1974), 39. See Robin Page Arnot, *South Wales Miners*, vol. 1, pp. 32–8 for the course of the strike.

53 *SWDN*, 12 August 1893. Daronwy Isaac was an associate of Mabon. He had been refused entrance to a miners' meeting the previous day.

54 Ness Edwards, *The History of the South Wales Miners* (London: The Labour Publishing Company Ltd, 1926), p. 115.

55 *TG*, 3 August 1893.

56 *Ibid.*

57 M. Severn, *The Miners' Evangel. A Text Book for all Manual Workers*, (Pontypridd:

John W. Ford, 1895), pp. 13–14.
58 *Ibid.*, p. 11.
59 *Ibid.*, p. 52.
60 *Ibid.*, p. 69.
61 *Ibid.*, p. 65.
62 *Regierungspräsident* Arnsberg to Minister of the Interior, 28 April 1896, STAD 9039.
63 The incident is described in Hue, *Die Bergarbeiter*, pp. 477–80.
64 O. Hue, *Neutrale oder parteiische Gewerkschaften? Ein Beitrage zur Gewerkschaftsfrage zugleich eine Geschichte der deutschen Bergarbeiterbewegung* (Bochum: Möller, 1900), p. 67.
65 *Ibid.*, pp. 93–4.
66 *Ibid.*, p. 83.
67 *Bk*, 29 April 1899.
68 Hue, *Die Bergarbeiter*, pp. 503–4. See also *Bk*, 12 and 22 July 1899 and *BZ*, 12 and 29 July 1899.
69 *EVZ*, 16 March 1897, clipping, STAD 9046.
70 G. Natorp, secretary of the *Bergbau Verein*, to a meeting of industrialists in Düsseldorf, 24 June 1878, cited in J. Paul, *Krupp*, p. 253.
71 For details on the cause of the strike and Brust's efforts to mediate see *Bk*, 1 May 1898.
72 *Bk*, 15 July 1898.
73 Hue, *Die Bergarbeiter*, pp. 491–500. Also expressed in O. Hue, *Mehr Bergarbeiterschutz* (Bochum: Möller, 1900), pp. 35–6. Brüggemeier adopts Hue's interpretation of the impact of the strike, *Leben vor Ort*, p. 198.
74 H. Imbusch, *Die Tätigkeit und Erfolge des Gewerkvereins christlicher Bergarbeiter Deutschlands* (Essen: Verlag des Gewerkvereins christlicher Bergarbeiter, 1915), pp. 11–12.
75 *Bk*, 15 July 1898.
76 *Protokoll der 16. Generalversammlung des Verband deutscher Bergarbeiter*, 10 to 15 June 1905, p. 45.
77 O. Hue, *Unsere Taktik beim Generalstreik* (Bochum: Hansmann & Co., 1905), p. 15.
78 *Regierungspräsident* Düsseldorf to Minster of the Interior, 30 January 1905, STAM OBA 1846.
79 See A. Gladen, 'Die Streiks der Bergarbeiter im Ruhrgebiet in den Jahren 1889, 1905 und 1912', in J. Reulecke (ed.), *Arbeiterbewegung am Rhein und Ruhr* (Wuppertal: Hammer, 1974), pp. 134–9; D. Fricke, *Der Ruhrbergarbeiterstreik von 1905* (Berlin: Rütten & Loening, 1955); Brüggemeier, *Leben vor Ort*, pp. 211–7.
80 G. Adelman, *Die soziale Betriebsverfassung des Ruhrbergbaus vom Anfang des 19. Jahrhunderts bis zum Ersten Weltkrieg unter besonderer Berücksichtigung des Industrie- und Handelskammerbezirks Essen* (Bonn: Ludwig Röhrscheid Verlag), 1962, pp. 122–5; H. G. Kirchoff, *Die Staatliche Sozialpolitik im Ruhrbergbau 1871–1914* (Cologne: Westdeutscher Verlag, 1957), pp. 151–9.
81 Arnot, *South Wales Miners*, vol. 1, pp. 45–60.
82 *SWDN*, 7 July 1898.
83 *CG*, 24 June 1898.
84 Pretty, 'David Morgan', p. 518.
85 *SWDN*, 6 July 1898.
86 *PC*, 24 June 1898.
87 MSWCOA meeting 1 September 1898, NLW MG 7.

88 *PC*, 9 September 1898.
89 *BZ*, 13 August 1904.
90 *Die Kämpfe um Knappschafts-Reformen im Allgemein Knappschaftsverein Bochum* (Bochum, 1910), p. 30.
91 *Bk*, 16 and 23 January 1904.
92 *Ibid.*, 6 August 1904.
93 See *Protokoll über die Verhandlungen des Bergarbeiter-Delegiertentagen für Preußen*, 28 March to 30 March 1905, p. 5, and *Protokoll über die Verhandlungen des II. Bergarbeiter-Delegiertentegen für Preußen*, 11 to 12 February 1906, pp. 20–22.
94 K. Tenfelde, 'Die Krawalle von Herne im Jahr 1899', *IWK*, 15 (1979), 71–104.
95 Hue, 'Die Krawalle von Herne', *Die Neue Zeit*, 43 (1898/99), 535.
96 *BZ*, 8 July 1899.
97 Kulczycki, *Foreign*, p. 93.
98 Reproduced in *Bk*, 5 April 1902.
99 On Polish subculture see Kleßmann, *Polnische Bergarbeiter im Ruhrgebiet*, pp. 83–92.
100 Kleßmann, *Polnische Bergarbeiter im Ruhrgebiet*, p. 120.
101 *Regierungspräsident* Düsseldorf to *Landrat* Essen, 15 December 1902, STAD 238.
102 Cited in Kleßmann, *Polnische Bergarbeiter im Ruhr*, p. 111.
103 Kulczycki, 'Trade union for Polish Miners', p. 615.
104 H. Imbusch, *Bergarbeiterstreik im Ruhrgebiet in Frühjahr 1912* (Cologne: Christl. Gewerkschaftsverlag, 1912), p. 44.
105 *Polizeiverwaltung* Hamborn to OBA 26 February 1905, STAM OBA 1845.
106 *Bezirks Polizeikommissar* Essen to *Regierungspräsident* Düsseldorf, 18 February 1905, STAM OBA 1846.
107 *BZ*, 28 October 1905.
108 *Geschäftsbericht des Vorstandes des Gewerkvereins christliche Bergarbeiter Deutschlands für die Jahre 1905 und 1906 und Protokoll der 11 Generalversammlung*, 29 June to 1 July 1907, pp. 42–3.
109 H. Imbusch, *Ist eine Verschmelzung der Bergarbeiterorganisationen möglich?* (Essen: Fredebeul, 1906), pp. 61–2.
110 Hue, *Neutrale oder parteiische Gewerkschaften*, pp. 8–9.
111 *Protokoll der 17. Generalversammlung des Verband der Bergarbeiter Deutschlands*, 5 to 9 May 1907, p. 60.
112 Hue, *Neutrale oder parteiische Gewerkschaften*, p. 114.
113 G. Werner, *Ein Kumpel. Erzählung aus dem Leben der Bergarbeiter* (Berlin, 1929), p. 149.
114 *Geschäftsbericht Verband der Bergarbeiter Deutschlands, 1911/1912*, p. 84.
115 *Regierungspräsident* Münster to Minister of the Interior, 5 February 1905, STAM OBA 1846.
116 Brüggemeier, *Leben vor Ort*, p. 199.
117 Berg, *Wirtshcaft und Gesellschaft*, p. 298.
118 See K. Schönhoven, *Expansion und Konzentration. Studien zur Entwicklung der Freien Gewerkschaften im Wilhelminischen Deutschland 1890 bis 1914* (Stuttgart: Klett-Cotta, 1980), pp. 264–9; A. Erdmann, *Die Christlichen Arbeiterbewegung in Deutschland* (Stuttgart: J. H. W Dietz, 1908), p. 451; M. Schneider, *Die Christlichen Gewerkschaften 1894–1933* (Bonn: Verlag Neue Gesellschaft, 1982), pp. 55–67.
119 Schönhoven, *Expansion und Konzentration*, p. 229.
120 Hue, *Neutrale oder parteiische Gewerkschaften*, p. 52.

121 Francis and Smith, *The Fed*, p. 19.
122 Blaenavon Lodge Minutes, 1 November 1906, MNA/NUM L/11/A3.
123 *Ibid.*, pp. 579–80.
124 Reports of the Manager of Locket's Merthyr Collieries Ltd to the Directors, 6 May and 8 June 1907, GRO D/D NCB16/35. T. I. 'Mardy' Jones was later elected as a Labour MP for Pontypridd.
125 The unions had supported the latter for some time. See for example, Otto Hue, *Mehr Bergarbeiterschutz*, p. 18.
126 See O. Hue, 'Arbeiterausschüsse als Arbeitervertretungen', and 'Die Arbeiterausschüsse in der Praxis', in *Sozialistische Monatshefte*, January and June 1907, Bd. 1, pp. 15–24 and pp. 489–63 respectively.
127 O. Hue, 'Weisse Salbe', in *Sozialistische Monatshefte*, April 1909, Bd. 1, p. 472. Later the *Gewerkverein* also claimed that employers placed pressure on safetymen to record '*Alles ist in Ordnung*' (everything is in order), *Bk*, 27 June 1914, clipping, STAM OBA 1853.
128 *Bk*, 3 February 1912.
129 H. Vogelsang, *Der Arbeiter-Ausschuss im Bergbau*, enclosed STAD OBA 1849.
130 *BZ*, 18 June 1910.
131 *Welche Aufgabe haben die Sicherheitsmänner?*, 1910, STAM OBA 1852.
132 *Oberbergamt* to Minister of the Interior, 2 January 1911, STAM OBA 1850 and *BZ*, 17 December 1910.
133 Adelman, *Betriebsverfassung*, pp. 145–9. See also *BZ*, 21 May 1914, clipping, STAM OBA 1853.
134 *Arbeiter-Zeitung*, 4 July 1910, clipping, STAM OBA 1850.
135 *RL*, 14 July 1900.
136 *RL*, 20 May 1905 and 9 July 1910.
137 Monthly Report of the No. 1 Rhondda District of the South Wales Miners' Federation, 1 May 1905, p. 27.
138 D. Hopkin, 'The great unrest in Wales 1910–13: A question of evidence', in Hopkin and Kealey (eds), *Class, Community and the Labour Movement*, pp. 249–75.
139 South Wales Miners' Federation Open Session on the Eight Hours' Bill, 17 March 1908, SWCC MNA/NUM/3/10/19.
140 H. M. Bock, 'Anarchosyndicalism in the German labour movement: A rediscovered minority tradition', in M. von der Linden and W. Thorpe (eds), *Revolutionary Syndicalism: An International Perspective* (Aldershot: Scolar Press, 1990), pp. 59–80 and J. White, 'Syndicalism in a mature industrial setting: The case of Britain', *ibid.*, pp. 101–18.
141 Bock, 'Anarchosyndicalism in the German labour movement', p. 62.
142 G. A. Ritter, *Arbeiterbewegung, Parteien und Parlamentarismus. Aufsätze zur deutschen Sozial- und Verfassungsgeschichte des 19. und 20. Jahrhunderts* (Göttingen: Vandenhoeck & Ruprecht, 1976), pp. 41–4.
143 White, 'Syndicalism in a mature industrial setting', pp. 101–18.
144 D. Egan, 'The Unofficial Reform Committee and the Miners' Next Step: Documents from the W. H. Mainwaring papers with an introduction and notes', *Llafur*, 2:1 (1978), 64–80. For the different viewpoints represented by two members of the URC see K. Davies, 'Rival Prophets? Williams Ferris Hay, Noah Ablett and the debate over working class political action in the south Wales coalfield 1910–1914', *Llafur*, 7:3/4 (1998/9), 89–100.
145 Egan, 'Unofficial Reform Committee', pp. 67–8.

146 *RL*, 9 July 1910.
147 Its initial membership amounted to just twenty, *Polizeikommissar* Gelsenkirchen to *Regierungspräsident* Münster, 11 February 1910, STAM Reg. Mü VII 82. By the end of 1910 the FVdG organised 450 miners, 250 of which came from the *Alter Verband*, K. Tenfelde, 'Linksradikale Strömungen in der Ruhrbergarbeiterschaft 1905 to 1919', in Mommsen and Borsdorf (eds), *Glück auf*, p. 216.
148 For the influence of the independent working-class movement upon the trade union activists in south Wales before World War One see Lewis, *Leaders and Teachers*, pp. 49–97.
149 *SWW*, 5 July 1913. See also R. Pitt, 'Educator and agitator: Charlie Gibbons 1888–1967', *Llafur*, 5:2 (1989), 72–83.
150 Francis and Smith, *The Fed*, p. 13.
151 Freie Vereinigung deutsches Gewerkschaften, *Das Programm der freien Vereinigung deutscher Gewerkschaften* (Berlin, 1906), pp. 10–15.
152 *The Industrial Syndicalist*, February 1911, p. 35. Italics in the original.
153 *The Miners' Next Step*, URC, 1912, reprinted with introduction from R. Merfyn Jones (Shoreditch: Pluto Press), p. 29.
154 *Ibid.*, p. 21.
155 *An die Bergsklaven*, enclosed *Bezirks Polzeikommissar* to *Regierungspräsident* Münster, 11 February 1910, STAM Regierung Münster VII 82.
156 *Bergarbeiter des Ruhrrevier*, enclosed *Bezirks Polzeikommissar* to *Regierungspräsident* Münster, 11 February 1910, STAM Regierung Münster VII 82.
157 *Ibid.*
158 *Step*, pp. 18–20.
159 *RS*, 1 March 1913.
160 Francis and Smith, *The Fed*, p. 22.
161 An example of this more radical, but not syndicalist, leader is Vernon Hartshorn. Peter Stead regards him as occupying an ideological position somewhere between Mabonism and syndicalism. P. Stead, 'Vernon Hartshorn: Miners' agent and cabinet minister', *Glamorgan Historian*, 6 (1969), 86.
162 Cited in Arnot, *South Wales Miners*, vol. 1, p. 322.
163 George Barker, miners' agent for Abertillery and member of the MFGB executive, directly linked the two strikes. See *RS*, March 1912.
164 *SWW*, 21 June 1913.
165 Sachse, *Protokoll der 19. Generalversammlung des Verband deustcher Bergarbeiter zu Bochum*, 1911, p. 89.
166 Brüggemeier, *Leben vor Ort*, p. 229. The *Dreibund* consisted of *Alter Verband*, ZZP and *Hirsch-Duncker* unions.
167 For examples of intimidation of working miners see STAM OBA 1857 and 1858.
168 See *BZ*, 30 March and *Bk*, 23 March 1912 respectively.
169 *Bk*, 2 March 1912.
170 Imbusch, *Bergarbeiterstreik im Ruhrgebiet*, p. 19.
171 *Ibid.*, pp. 32–7.
172 *Ibid.*, p. 23.
173 *BZ*, 10 February 1912.
174 Huseman, *Protokoll der 20. Generalversammlung des Verband deutscher Bergarbeiter zu Hannover*, 1912, p. 103.
175 R. Roberts, *A Ragged Schooling: Growing Up in a Classic Slum* (Manchester: Manchester University Press, 1976), pp. 195–6.

# 3

# Political languages, 1890–1914

Devotion to Liberalism is a Welsh national trait; it is inherent in the Welsh national character.

*South Wales Daily News*, 12 January 1906.[1]

. . . still in the hands of the clerics, who, at every opportunity, more or less denounce the Social Democrats and the free trade unions as Beelzebub and his followers . . . So that today the political as well as the trade union movement is still, as for ten or twenty years, almost exclusively sustained from the foreign elements.

The *Arbeiter-Sekretariat* (workers' secretariat) Bochum commenting on the SPD and free trade union movement in the area, 1905.[2]

## Introduction

The pattern of homogenisation and fragmentation identified in the examination of the miners' trade unions had parallels within the political sphere. While the parliamentary representation of the south Wales miners remained largely the preserve of the Liberal Party up to 1908, there was fierce competition in the Ruhr between the German Centre Party, the National Liberals and, later, the Social Democratic Party for the votes of the miners. Therefore, the dual trends of homogenisation and fragmentation also applied to the political as well as the industrial arena.

As with the trade unions, these trends were reinforced by the identities created by the various political parties. As such this chapter poses similar questions to its predecessor. How did the parties construct a discursive identity with which to appeal to the miners? What were the constitutive features of these identities? How did they relate to the milieux highlighted in Chapter 1? On the one hand, the chapter will focus on the Liberal discourse in south Wales and examine how the party sought to deal with the growing concern with labour as a political entity. On the other, it will question why the SPD, despite the overwhelmingly industrial character of the region, was ultimately unable to turn the Ruhr into a political stronghold.

Before examining the discourses of the various parties some attention must be given to the institutional background. As shown in the previous chapter institutional factors both limited and offered opportunities for the expression of discourse. Given the different electoral systems of Britain and Germany, this was even more the case in the political sphere. These systems provided the framework within which discourse was played out. Furthermore, between the individual acceptance of a particular discourse and the casting of the vote there lies a whole gamut of intermediate factors. Results can be, and were, massaged and distorted by gerrymandering, ballot stuffing, coercion, treating, plural voting, limited or unequal suffrage, the banning of party meetings, and agreements between parties to 'deliver' their electorate. The first section therefore briefly considers the implication that the electoral systems had for the transmission of discourse.

## The political systems of Germany and Britain

In both Germany and Britain 1867 represented something of a political watershed. In Britain Disraeli's 1867 Reform Act expanded the electorate from 1.4 million to 2.52 million. Later the 1884 Franchise Act introduced by the Gladstone government extended the vote to every householder in the counties, increasing the electorate to 5 million. In Prussia there was a more dramatic change. Bismarck introduced universal male suffrage for the Reichstag elections, a provision that was adopted in the Reich Constitution after the Franco-Prussian War. In both cases it was pragmatic political considerations that motivated the legislators to extend the vote. Disraeli needed the support of radical Liberals to keep his minority Conservative government in power, while Bismarck sought to out-manoeuvre the German Liberals, believing that the elections would provide a Conservative majority.[3]

Apart from the different size of their electorates there were other significant differences between Germany and Britain. Not only did more Germans enjoy the vote; they also had greater opportunity to vote.[4] The federal structure of the German Reich meant that there was another representative tier between the local and national levels. However, universal male suffrage did not apply to the Prussian *Landtag*, within whose boundaries the Ruhr coalfield lay. The electorate for the *Landtag* was divided into three classes according to the rate of tax paid, a pattern also applied to local elections. Under this system each class elected a third of the representatives, a feature that worked to the advantage of the industrialists. In industrial communities of the Ruhr the mineowners, as the highest tax payers, were often the only voters in the first class, allowing them to determine a third of representatives directly.[5]

Britain lacked an intermediate elective layer between the local and national. Furthermore, until the passage of the 1888 Local Government Act, the unelected Justices of the Peace and piecemeal sanitary authorities governed the counties. The Local Government Act established elected county councils. The counties were further sub-divided into rural and urban district councils, although different franchises obtained in different areas. Finally, while voting in Britain was on a 'first past the post' basis, in Germany, should no one achieve an overall majority, a second run-off round would be held between the two leading candidates.

There were two important consequences of electoral systems for party political development and discourse. First, the run-off system in Germany encouraged a plurality of parties. Should a party lose the first run they could also strike a deal with one of the parties going into run-off, delivering their vote in return for concession.[6] In this way Germany's electoral system promoted a multiplicity of parties and, since every party needed to distinguish itself from its opponents, of discourses. On the other hand, the first-past-the-post system in Britain did not offer these opportunities. Instead it was more advantageous to build up alliances of interests prior to the election as division could cost a party victory. Therefore, several interests were contained within the 'broad church' of one of the two main parties for much of the late nineteenth century. For example, the Newcastle Programme, produced by the National Liberal conference in October 1891 and adopted by Gladstone, included a wide range of reforms covering temperance, agriculture, religion and labour issues. It has been argued that the varied nature of this programme revealed the sectarian divisions within the party itself, of which the Welsh group was but one.[7] On the other hand, the Welsh contingent has been regarded as a vigorous and positive force within the party, able to unite 'Old Liberal' concerns with religious issues and the focus of 'New Liberalism' upon social and labour questions.[8]

Second, universal male suffrage in Germany meant that a self-declared working-class party had at least a chance of winning votes. There existed a specific electorate at which they could aim their appeal.[9] The disenfranchisement of a large section of the British working class meant that, during the nineteenth century, labour interests were more often expressed through the existing parties. The clearest expression of this was the emergence of the so-called 'Lib-Lab' MPs, who combined traditional Liberal concerns with an interest in policies directly affecting labour. Mabon, the leading Lib-Lab MP in the south Wales coalfield following his election in 1885, summed up this point of view by claiming that, 'I am pledged to vote for Labour reform, whatever Government brings in the Bill, but as an individual I am a Liberal.'[10] These institutional and structural factors contribute to explaining why, in Britain, the Liberals were largely able to accommodate labour demands up until the twentieth century, while specifically working-class parties were formed in Germany at a comparatively early date.[11]

## Political discourses, 1890–1900

This examination of the discourses of the various political parties mainly focuses on several key constituencies and individuals that were paradigmatic of discursive trends. For the Ruhr, the electoral districts are Dortmund-Hörde, Bochum-Gelsenkirchen and Essen. The SPD ultimately managed to win

Dortmund from the National Liberals and turn it into an SPD stronghold. Essen, on the other hand, remained firmly under the control of the Centre Party throughout the *Kaiserreich*, except for the period between 1893 and 1898. Finally, Bochum swapped hands between all the parties, but by the late 1890s it had become a battleground for the SPD and National Liberals. In south Wales the constituencies are the Merthyr Boroughs, Glamorgan (Rhondda) and Glamorgan East. The Merthyr Boroughs had a reputation for radical politics. The Rhondda was held by that epitome of 'Lib-Labism', Mabon, and East Glamorgan repeatedly returned the Liberal Alfred Thomas for much of this period.

As already noted the real breakthrough for the SPD in the Ruhr came in 1890. Prior to the lapse of the Anti-Socialist Law its political importance was negligible in the area. Once permitted to campaign freely its percentage share of the vote rose steeply from 2.5 per cent in 1887 to 11.5 per cent in 1890.[12] Unsurprisingly, the issue of class featured strongly in SPD discourse. In much of the Ruhr, and in Dortmund in particular, one of the party's primary means of disseminating its discourse was through its organ *Die Rheinische-Westfälische Arbeiterzeitung* (*RWAZ*). The paper sought to construct a dichotomous identity based on the perceived conflict of interests between the working class and the employers. In 1893 Carl Wilhelm Tölcke, a long-standing member and influential figure within the social-democratic movement, opposed the incumbent National Liberal candidate, Theodor Adolf von Möller, a factory owner and co-founder of the Central Association of German Industry (*Centralverbandes deutscher Industrieller*, CdI). During the campaign a series of articles appeared under the title '*Volk der Arbeit aufgewacht*' (People of Work Awake). Within these articles the National Liberals were portrayed as the party of the capitalist owners, whose interests were necessarily opposed to those of the working class.

> What is a party? A party is an interest group, in other words: people who have the same interests towards the economic and political institutions of the state join together in a party. Now since the interests of the factory owners, the coal barons, the big businessmen, etc. of all exploiters in general, are diametrically opposed to the interests of the workers, it is clear that every worker who votes for a candidate of the capitalist parties commits an act of treason against his own interests.[13]

Möller himself was characterised as a representative of 'mobile capital'. On the other hand, a vote for the SPD was seen as a 'protest against the prevailing system of exploitation'.[14] By implication Tölcke and the SPD represented the true interests of the working class. Only by voting for the SPD would a miner be faithful to his own interests, which were indivisible from those of the working class as a whole.

However, the SPD in Dortmund, as elsewhere in the Ruhr, was fighting on two fronts. The Centre Party, although weak in Dortmund, still offered a challenge to the SPD. Its reluctant candidate in 1893 was Lambert Lensing, founder and editor of the Catholic organ *Tremonia*.[15] In its 'othering' of the Centre the SPD took a different tack. A discursive construction of the Centre as merely an

*Table 3.1* Estimated party strength in Dortmund by class (by percentage)

| Class | Socialists | National Liberals | Centre |
|---|---|---|---|
| White-collar occupations | 21 | 91 | 44 |
| Blue-collar occupations | 79 | 9 | 55 |
| Total | 100 | 100 | 99 |

*Source:* Stanley Suval, *Electoral Politics in Wilhelmine Germany* (London: University of North Carolina Press, 1985), p. 83.

employers' party was not entirely convincing, since it received roughly equal support from both blue and white-collar groups.

The Centre Party, after all, had links to a dense network of workers' organisations in the form of the Catholic *Arbeitervereine.* Furthermore, the Centre itself drew distinctions between the Catholic working class and the predominantly Protestant owners. Therefore, rather than portraying the Centre as solely a party of the employers, the SPD laid more emphasis upon the way it split the miners' movement. Thus, in the same address in which he condemned Möller as the puppet of the employers, Tölcke blamed the Centre for weakening working-class interests by dividing the movement against itself.[16]

However, class identities were also imposed on the Centre Party by the SPD. An election declaration that appeared in the paper on the day of the election called on the workers to reflect upon the interests that the various parties represented. While the National Liberals represented the 'chimney and mine barons', the Centre represented the 'propertied middle class, or rich aristocrats and well-paid Church officials; they will therefore only represent their interests, which are in opposition to those of the working class'. In some respects the Centre was seen as being made up of feudal and backward elements, but these were just as intent on exploiting the workers as the industrialists. Finally, the small Freisinnige and Democratic parties were characterised as representatives of the lower middle class (*Kleinbürgertum*), who could only impotently dream of power. On the other hand, the:

> social-democratic party consists of workers, is the party of the worker. Social Democracy must therefore represent the interests of the workers in contrast to all the other parties of the propertied middle-classes. The division is sharply drawn. On the one hand stand the propertied classes, to be sure split between various conflicts of interest, but completely united against the 'covetous' workers, united so that the present exploitation of the workers through the capitalist employers can be maintained. On the other side stand the wage workers (*Lohnarbeiter*) as an independent class, represented by Social Democracy, which will bring their exploitation by the capitalist employers to an end.[17]

Thus, while SPD discourse had certain nuances and established links between various opposition parties and certain socio-economic groups in Wilhelmine society, it ultimately conflated these differences to produce an essential dichotomy between the 'propertied' and the workers. The SPD regarded the political sphere as a direct extension of the social sphere. The fundamental conflict between capital and labour was reflected in the political parties. There existed no autonomous space for politics. This discursive interpretation of society was entirely in keeping with the Marxist theory that the party had accepted as its ideological basis at the 1891 Erfurt conference,[18] and propagated through both its more radical (*Die Neue Zeit*) and reformist journals (*Sozialistische Monatshefte*). The experience of state repression through the Anti-Socialist Laws was instrumental in encouraging a whole generation of Social Democrats to endorse Marxism. Although Lassallean ideas remained an undercurrent within the party, the essential conflict at the root of Marxist ideology seemed to correlate better to the discrimination the Social Democrats experienced under Bismarck.[19] The Anti-Socialist Laws also reinforced the earlier split between the labour and liberal movement in Germany. The SPD view that religion was an obstacle to full working-class solidarity was closely bound up with this acceptance of Marxism. For the SPD class solidarity should be the sole interest of the workers. This 'totalising' conception of socialism rigidly defined the contours of SPD discourse, limiting its ability and willingness to work with the Christian-Socialists.

The labour–capital dichotomy remained the core theme in SPD discourse throughout this period. Admittedly, there were internal differences within the SPD and three groups can be distinguished. The orthodox Marxists, represented by Karl Kautsky, awaited the revolution that would be the inevitable result of capitalism. The *Jungen*, on the other hand, sought to precipitate the revolution. Finally, the revisionists rejected revolutionary tactics in favour of practical reform within the existing system.[20] It was ultimately this latter group that began to gain the upper hand in the movement before the war. Yet, despite these differences, the perception that the interests of workers and owners were essentially in conflict was common to all.

In subsequent elections in Dortmund the *RWAZ* used the same kind of rhetoric as it had in 1893. In a by-election in October 1895 the SPD was actually successful in returning its deputy for Dortmund. However, Dr Lütgenau's hold on the constituency was tenuous. During the 1898 elections the paper complained bitterly that:

> numerous workers have still not yet awoken to class consciousness. That means they have not yet clearly recognised the conflict of interest between the great owners, the small owners and the proletariat and they have therefore not gained the insight that [*Reichstag*] deputies, which belong to the owners' parties, could not possibly observe both the interests of the workers and the middle-classes at the same time, since the interests of both sides are in opposition.

Furthermore, the article called upon the workers to look to what the party stood for rather than concentrating on the personality of the candidate.[21] Here again working-class interests, notions of class consciousness and voting for the SPD were conflated into one and the same thing. Only by voting SPD could the miner express the fact that he had fully realised his true identity as a member of the working class. Ultimately, the Dortmund party's fears over the underdeveloped sense of class consciousness in the constituency seemed to have been realised as Lütgenau lost the seat to the National Liberal candidate Alexander Hilbck in the run-offs. However, the paper placed much of the blame upon 'the surprise of the Centre electorate by the leaders', who would rather throw their support behind the Protestant, possibly anti-Catholic, Hilbck than let an irreligious Social Democrat win.[22]

Parallel discourses to that of the SPD are difficult to discern in south Wales during the 1890s. Britain as a whole lacked a large, self-consciously working-class party on the same lines as the SPD. Admittedly, the Social Democratic Federation and Socialist League had been active in south Wales in the 1880s. In the late 1890s the Independent Labour Party (ILP) also sought to extend its organisation into the principality. These associations carried out agitation tours on the back of the industrial unrest that disturbed the region during that period.[23] However, membership remained small and they did not stand for elections for much of this period.[24]

Common motifs can, however, be identified within the discursive constructions of the British Liberals, National Liberals and Centre. One such common theme was that of religion. All three parties used this as a basis upon which to build a political identity. In south Wales, the Liberal Party was widely regarded as the party of Nonconformity and disestablishment was one of the key planks in its platform. The discourse of Alfred Thomas, Liberal MP for East Glamorgan, was illustrative of the important role Nonconformity played within Liberal discourse in south Wales. At a rally in support of Alfred Thomas's candidature held at Pontypridd during the 1885 election, Thomas Williams, President of the East Glamorgan Three Hundred, challenged the audience to be true to their beliefs: 'If you are a Conservative, vote for the Tory; but if [you are] Liberals and Nonconformists vote for Mr Alfred Thomas as the representative of your principles.'[25] In subsequent elections, all of which Thomas won, the same Anglican-Tory/Nonconformist-Liberal dichotomy was deployed. Ten years after he was first elected, during the 1895 elections, Thomas faced a challenge from the Conservative, Henry Jackson. One musically inclined Liberal supporter placed the discursively constructed identity into song.[26]

Should old acquaintance be forgot
And thrown away for new
What good would Jackson do for us
That Thomas will not do

*Chorus:* For Thomas is man of men
A Liberal good and true
A Cymro pure we'll send again
Our battle to renew

Where Lewis failed shall Jackson win?
As Tory democrat
No! East Glamorgan answers plain
There is no fear of that

For Jackson stands for privilege
For wealth he stands, and class;
But Thomas stands for liberty
The welfare of the mass

Jackson would keep an Alien Church
To weigh our nation down
But Thomas means equality
Our freedom yet to crown!

Say what have Tories done for us
Where should we stand to-day
If Tories and their Alien Church
Had their own sweet way

Again would liberty be dumb
Again would hide her head
No never will we yield the faith
For which our fathers bled

For Liberty means no man's wrong
Its mission is to bless
To shield the helpless from the strong
To succour all distress

If East Glamorgan but unite
As she had done before
Then Thomas need not fear to fight
And Jackson fight no more.

Its musical merits aside, this election song neatly summed up some of the main features of Liberal discourse. Again the Tories were linked to the interests of the Anglican Church, the song carrying the implication that the Liberals would naturally represent Nonconformity. However, to the religious imagery was added a nationalist tone. By 'othering' the Anglican Church as 'alien', the Liberal discourse depicted it as something foreign to the Welsh 'nation' and its people. The intermingling of nationalism and religion was furthered in the sixth verse by reaching back to a supposed Welsh history, in which Nonconformist forefathers made sacrifices for their faith. This dichotomy between Welsh/ Nonconformist and English/Anglican was capped by the description of Thomas as a 'Cymro'. Thus Thomas was represented as the natural choice for the Welsh-Nonconformist voter, while Jackson was characterised as the unnatural, 'foreign' candidate. Within this discourse Nonconformity was being linked not only to Liberalism, but also to Welsh national identity.[27] True Welshmen were Nonconformists and vice versa. A distinction was drawn between these values and Toryism, which represented their negation. Voters could give physical manifestation to their Welsh identity by the relatively simple expedient of voting Liberal.[28]

Kenneth O. Morgan has argued that around 1892/3 there was a divergence between Liberalism and nationalism and points to the failure of the Cymru Fydd movement to find widespread support.[29] However, the issue of Welsh identity continued to be used within Liberal discourse as a means of mobilising the voters against their Conservative supporters. The combination of religion and nationalism played an important role in the myth of the *gwerin*, which can be roughly equated with the German conception of the *Volk*. Within this conception the Welsh were idealised as 'a cultivated, educated, often self-educated, responsible, self-disciplined, respectable but on the whole genially poor or perhaps small-propertied people, straddling groups perceived as classes in other, less fortunate societies'.[30] This independent, dissenting and generally Welsh-speaking populace had a rural character. Neil Evans and Kate Sullivan have shown how the Liberal Party used the Welsh language itself as a means of fostering this identity and as a weapon against the Tories when campaigning.[31] The Welsh peasantry was the ideal of the *gwerinol* discourse and in Liberal discourse they were placed in juxtaposition to the Anglicised, Church of England supporting, landowners, whose interests were seen as embodied by the Tory Party.

Indeed, although miners dominated the East Glamorganshire constituency by the end of the nineteenth century, Thomas continued to appeal to the *gwerin* ideal. At a meeting at Cilfynydd Thomas's speech contained three main points: Irish Home Rule, Disestablishment and the Welsh Land Bill.[32] Similarly, during the 1900 campaign Thomas moved seamlessly from religious to agricultural issues when criticising the achievements of the Tory Government.

> They had the Conscience Clause, but, of all frauds, there was no greater fraud than that. How could they expect a little child to say 'I am a Nonconformist' and refuse to join the catechism, and then have the finger of scorn pointed at him? He was surprised at the Nonconformist parents (suffering as they had) allowing their children to be taught by ministers who did not belong to the Church of their fathers. The Conservatives had also remembered the landlords. They had taken off half the Poor Rates from the tenants. Had not the tenants had the advantage? Before that had come about they had heard of such-and-such a landlord giving back 20, 30 or 40 per cent to their tenants; but now that that law had been passed they never hear of that kind of thing. Who has benefited by that law? The landlords. The poor man who bought tea at 1s 4d per pound paid the same duty as the rich man who paid four or five times the amount for his tea. That was only one small matter, but it showed the principle.[33]

The extent to which this nationalist, *gwerinol* discourse was used by the Liberals of south Wales can be illustrated by a brief comparison with the political discourse of one of the most successful Conservative politicians in the region. Colonel Frederic Courtney Morgan, son of Viscount Tredegar, represented the South Monmouthshire constituency from 1874 until 1906. Within his election speeches there is virtually no mention of Wales as a distinct nation. In fact, following his 1880 victory he even declared that 'the electors of the county had taught the rest of England a great lesson, stick to Conservatism'.[34] His only reference to Wales seems to have been as a result of a direct question during the 1895 election campaign. When asked if he was a 'Welsh member [of Parliament]' he replied, 'No I am not. I am a Welshman, but I was the member for Monmouthshire, or a portion of it.'[35] Admittedly, Monmouthshire was the most Anglicised of the Welsh constituencies and was not necessarily considered part of Wales. But the complete lack of any sense of Welsh nationalism is striking when considered alongside Alfred Thomas's discourse. Instead, Morgan, along with other Conservative candidates for the Monmouthshire seats, focused upon a sense of 'Britishness' and the Empire. In 1900 he elided imperial and welfarist notions when he pledged 'to uphold the integrity of the Empire, the welfare of the people, and the interests of the district I represent'.[36] Morgan's counterpart in the North Monmouthshire constituency, de F. Pennefather, also sought to link the interests of Empire to those of the workers. He claimed 'as a practical man he knew the greatness of the Empire depended upon the workers'.[37]

Although Morgan's lack of concern with 'Welshness' may be explained by the ambiguous position of his constituency, the same rejection of a Welsh national identity in favour of a British one is found elsewhere. Conservatives at the opposite end of the coalfield also emphasised the Empire and Britain over Wales. In 1906 the Conservative candidate for Carmarthen was a Mr Ponsonby. He campaigned almost exclusively on the issue of Tariff Reform, stressing the benefits duties would bring the Empire and the working class, but did not mention Wales once.[38] Similarly, four years later a whole page in the *Carmarthen Journal* was devoted to a Conservative election notice in support of Mervyn Peel for East Carmarthen and Lord Tiverton for Carmarthen Boroughs. It consisted of four questions, each of which began 'Are you, as Britishers?'[39] The contrast is even more striking when placed alongside Liberal discourse in the constituency. The Carmarthenshire Liberals played extensively on their Welsh credentials. The incumbent, Major Jones, declared in 1895 that the 'question for them was not what would happen in Manchester or Derby, but what was to happen in Wales'.[40] Fifteen years later Liberal supporters claimed that it was their ideal 'all along that they in Wales at any rate should get men who were in entire sympathy with their aspirations as a country and as a nation. Their ideals and aspiration were national, and characteristic of themselves as a nation, and they liked to get a man who was one of them'.[41]

It is notable that there is little mention of class in any of the above examples of Liberal or Conservative discourse. In fact, there is scant mention of the working class in any of Alfred Thomas's speeches during this period. Where class is mentioned, as in Thomas's election song, it is imbued with negative connotations and refers to the landowners. The contrast between the 'classes' and the 'masses' implies that the former represent sectional interests, while the concept of the *gwerin* was elided with the 'masses'. In fact, in much of Liberal discourse there was a dichotomy between 'the people' and the landowners, rather than between classes as in SPD discourse. In 1895 an article appeared in the *Glamorgan Free Press* supporting the candidatures of Thomas, for East Glamorgan, and Arthur J. Williams, for South Glamorgan. It was comprised of three sections appealing to farmers, workers and Nonconformists respectively. The section on the workers concentrated upon the Liberals as the party of labour reform. However, in the whole piece there was no mention of class. Instead, emphasis was placed upon 'the people' and their opposites were once again the landowners rather than the employers.

> If you want men to represent the landlords and the landlords' institution vote for Major Quinn and Mr Jackson [the Conservative candidates], but if you want men who will vindicate the rights of the House of Commons as the mouthpiece of the people's wishes and claims, do not forget to vote for Messers Arthur Williams and Alfred Thomas, and the truth of the old adage, Trech gwlad nag arglwydd.[42]

The tone of the piece is inclusive, attempting to unite farmers, workers and Nonconformists against a perceived landowning oligarchy.[43]

Within Liberal discourse then the workers were subsumed within this conception of the people. In effect Liberal discourse sought to transcend class. If classes were recognised at all it was in a cooperative sense.[44] Differences between certain groups were recognised, but, just as the SPD did with the German upper and middle classes, these were conflated to present the people as an essentially homogeneous group. In doing so the Liberals presented a communitarian ethos. Their MPs claimed to represent communities as a whole, rather than certain occupational groups. Therefore, the 'politics of personality' suited Liberal discourse as it emphasised notable individuals from within the community.[45] Thus, candidates often emphasised their local origins or connections. When D. A. Thomas stood for Merthyr in the March 1888 by-election he was quick to point out that 'he was a native of this borough, and a Welshman on both sides (cheers). In addition to that he was closely associated with one of the staple industries of the district and, as his prosperity depended to a great extent upon theirs, he was likely to safeguard their interest in the House of Commons.' Welsh identity and Liberal credentials again served as the crux of Liberal discourse. He spoke 'in the vernacular' to convince his audience of his knowledge of Welsh and sharply attacked the Church. 'The people were alive to the fact that their national property, in the shape of tithes, was going into the coffers of an establishment entirely out of sympathy with their sentiments and feelings.'[46] The recurrent theme of nationalism and religion aside, it is notable that the fact that D. A. Thomas was a coalowner was not of prime importance. What was important was the implied interdependent relationship between his works and the miners. The two were indivisible and by electing D. A. Thomas the miners would have an MP who represented the whole community.

A similar combination of community, religion and nationalism was evident in the discourse of the National Liberal and Centre parties in Germany. However, while the British Liberal Party in Wales was able to articulate a specific nationalism within Britain, the National Liberals and Centre parties' discourse was concerned with placing themselves within the context of a German identity. The National Liberals continually played on this nationalist motif. The election declaration for the 1890 campaign called upon the voters to ensure that a government-friendly majority was elected to the Reichstag.

> To ensure the election of such a majority is today more than the duty of all friends of the Fatherland. Not in the interest of the party, but for the Fatherland we call to our friends that everyone does his duty. It is yours, and the affairs of the German Empire, which are dealt with. Allow yourselves to be led by the great national point of view, not the differences of opinion and conflicts of interest.[47]

Specific issues such as increased military expenditure (*Militärvorlage* ) provided a basis for the nationalist propaganda of the National Liberals.[48] At a rally on 4 June 1893 the National Liberal candidate for Bochum Dr Haarmann, the Bürgermeister of Witten, concentrated on the dangerous position that Germany was in and appealed to patriotism. To undermine the SPD and the Centre Party he pointed out:

> If the speakers of the opposition claim that France does not think about war, then the question must be put, why is [France] continually and to the extreme arming itself. France has nothing to fear from Germany. Only thoughts of revenge, already fostered in the school, form the driving force. The opponents point out the high costs, which acceptance of the submission will bring. Admittedly the costs are high, but they are not prohibitive and they must be affordable, if we can only turn away the threat of war or be victorious through a strengthening of the army.[49]

By focusing on the danger posed by France Haarmann was also reaching back to the very foundations of the German Empire and able to play on feelings of national pride. Furthermore, he was able to 'other' the opposition by presenting their refusal to support the bill as at least misguided, if not unpatriotic. Haarmann was especially critical towards the Centre as they too sought to appropriate nationalism as a feature of their discourse. 'They always speak as if no sacrifice is too great for the honour and the greatness of the Fatherland and when the time arrives, then they refuse the means for the execution of the *Militärvorlage* necessary for Germany's security.'[50] Similarly, Möller in Dortmund appealed to the voters 'on election day only to give their vote to a patriotic thinking and feeling man, a man, who is primarily concerned with upholding the security of the Fatherland'.[51] Thus within National Liberal discourse identification with government policy often constituted the measure of one's patriotism. This notion of German identity was a recurrent theme in National Liberal discourse throughout the *Kaiserreich*.

However, the National Liberals were unable to appropriate this entirely. A multi-polar ideology, nationalism could be filled with a variety of meanings. Thus, the Centre Party was also able to claim the mantle of nationalism. For Centre Party activists, however, this could encompass a more critical stance towards government. Within Centre discourse 'true patriotism is not shown by saying yes and amen to all government proposals, but that we are of the opinion that the government must listen to the people'.[52] This allowed them the latitude to attack certain policies such as the the *Militärvorlage* in a leaflet, which it declared would be injurious to the interests of artisans.[53]

At the same time, however, the party was sensitive to accusations of being un-German and sought constantly to reaffirm its loyalty to the *Kaiser*. The combination of monarchism and Catholic sentiment evident in the poem 'Pope and Kaiser' was typical of Centre discourse.[54]

Holy Father, to you,
Leo, we devote to you
The devotional greeting!
For you we sing with delight
Always from happy breast,
We think about you today,
Hail, to you

Hail to you in the crown,
Ruler of the Fatherland,
Hail to you Kaiser!
Feel in the throne,
The lofty bliss,
Of being the darling of the people
Hail, to you, Kaiser

Hear in Heaven high,
Lord, the supplication of your people
And bless them,
That they with wise hand,
Church and Fatherland,
In unity lead
To tranquillity

Both the National Liberal and the Centre parties used this nationalist discourse to attack the SPD. For both the SPD represented the party of revolution and internationalism, threatening the established social order. It was due to this shared perception that, in many of the constituencies in the Ruhr, alliances, sometimes formal, sometimes informal, were formed to keep the SPD out. A prime example was the 1898 Dortmund election. The local Centre Party, having lost the first round, advised its members to vote for the National Liberal candidate, Hilbck, in the run-offs, in order to unseat the SPD incumbent Dr

Lütgenau.[55] Interestingly, and a reflection of the party's knowledge of their position in the constituency, *Tremonia* had been advocating support for Hilbck in the run-offs before the main election had actually taken place. A table was even printed mapping the various options for the run-offs depending on the results of the main election.[56] Furthermore, the Centre still supported Hilbck against Lütgenau, despite complaining that some mining officials had tried to influence their workers during the first round.[57] The legitimisation for supporting Hilbck, despite these alleged underhanded tactics, was the SPD's revolutionary nature. 'Social Democracy is not only the most dangerous nemesis of the Church and the Christian family, but also the existing social and civil order.'[58]

In fact Hilbck had deliberately sought to exploit the religious and nationalist concerns within both National Liberal and Centre discourse by adopting a broad approach. Throughout his campaign Hilbck had sought to depict the National Liberals as a *Volkspartei*, a people's party. His election declaration claimed that he would represent Protestants, Catholics, industry, agriculture, tradesmen and artisans, officials and workers all under the black-white-red flag.[59] At a rally in Lütgendortmund he declared his hope that 'this party (the Centre) will gradually develop from an ultramontan into a national [party]'. This was in sharp contrast to his characterisation of the SPD, which would always be un-national. 'It had been proved how the daze of internationalism haunts only the head of the German Social Democrats, while the socialists of other lands always stand on the side of their country in national questions.'[60] During the run-offs he played this nationalist card to its full extent to draw out the Centre voters, claiming 'no patriotic man can be in doubt over what he has to do'.[61]

Krupp too took a non-confessional stance in his 1893 campaign in Essen, previously a Catholic stronghold. The Krupp campaign represents the best example of how the politics of personality functioned in the Ruhr. This was especially so since Krupp himself did not deign to involve himself in the campaign. His sole contribution appears to have been a letter published in the *RWZ*. In this he concentrated on the fact that, as an employer of both Catholic and Protestant workers, he was not interested in any religious conflict. 'I demand, as my father did, in the interest of my works and its affiliates, that Protestant and Catholics work together in friendship.'[62] Supporters of Krupp's candidature in Essen furthered this inclusive tone.

> Catholic voters can also vote without reservation for the candidate of the national parties, Herr Krupp, since Herr Krupp, although Protestant would likewise represent the interests of the Catholic citizens as well as the Protestants ... The name Krupp has such an outstanding importance for Essen and the area there could be no more natural or better representative than Herr Krupp.

Much like D. A. Thomas in the Merthyr constituency, Krupp's interests were portrayed as indivisible from the interests of the community as a whole. Finally, as for candidates in other Ruhr seats, the *Militärvorlage* was used to add a nationalist tone. In this case it was presented as something above party politics, necessary for the defence of Germany.[63]

However, religious differences, and the related issues of national identity, could often place the National Liberals and Centre in sharp conflict. This was the case in Bochum in the 1890s. Just before the 1893 run-off the Centre had produced a leaflet apparently signed by leading local Social Democrats, which advised their supporters to vote for the Centre candidate Fuchs. The *RWAZ* later denied this as a forgery, but only after Fuchs had won.[64] The 1898 election was marked by particularly robust discourse based on religion as the National Liberals sought to mobilise the Protestant community in Bochum against the Centre. Just before the run-off election the Protestant pastors of the district produced a pamphlet aimed at the faithful. Once again Protestantism was elided with nationalism and fear of the enemy within. Protestant voters were asked whether they wanted their 'great industrial district [to] be represented by a foreign, Cologne merchant, who belongs to the great, ultramontan party which would like to place Germany again under the crooked rod (*Krummstab*) of the Roman Pope'. The pamphlet ended with the obligatory call to duty for the Protestant, German man.[65]

The Centre, meanwhile, invoked the *Kulturkampf* in support of Fuchs. At a meeting on 2 June 1898 it was claimed 'the *Kulturkampf* is finished, but its advocates (*Kulturkämpfer*) wait only for the chance when the Centre loses its crucial power to give undisguised expression to their feelings'.[66] Thus, while the National Liberals tended to focus upon monarchist sentiments and the memory of the creation of Germany, the Centre were able to play upon fears that a renewed *Kulturkampf* could be just around the corner.

During the 1898 election campaign in Bochum it was also claimed that Franken paid low wages at his factory, an accusation that was refuted by his Catholic workers in a counter leaflet.[67] This concern with presenting themselves as interested in the welfare of the workers was common to the south Wales Liberals, the National Liberals and the Centre. In 1893 the Fuchs campaign had produced a pamphlet reproducing a speech the politician made regarding an accident at the *Carolinenglück* mine in which he supported workers' councils and celebrated the victims as 'knights of work, fallen on the field of honour'.[68] In 1898 the National Liberals adopted a similar stance and stressed that he was the workers' friend. Fuchs had done nothing for the miners but, 'Herr Franken, who, as a workers' child, grew up in work, is no friend of hollow phrases, but a man of action'.[69]

This 'friend of the worker' theme was common to the British Liberals, National Liberals and Centre. It was implicit in the importance attached to representing the community, since it was believed that candidates, who were often industrialists or businessmen, would by pursuing their own interests naturally produce benefits for their employees. However, within British Liberal and Centre Party discourses currents existed which identified more directly with the working class. Two prime examples are Mabon in the Rhondda and Gerhard Stötzel in Essen. Both came from working-class backgrounds and were elected as workers' candidates in opposition to middle-class opponents from their own parties.[70] In their subsequent election campaigns neither was shy of using this identification to their advantage.

In the case of Stötzel this was particularly evident in 1890 when he faced a challenge to his monopoly of working-class identity in the shape of Ludwig Schröder, *Alter Verband* leader, miners' delegate to the *Kaiser* and SPD candidate. Stötzel's working-class credentials were continually emphasised at rallies. Stötzel was presented as a man who knew:

> the circumstances of the workers as well as the best miner. In Bochum they have selected a miners' candidate, but there are also other interests to be represented in the Reichstag. We have also iron and steelworkers, who could also use the same right and name a candidate from their midst. We would be making a great mistake if we separate ourselves from Herr Stötzel and are not loyal to him because he is a strong-minded man respected by all parties.[71]

In a foreshadowing of the differences that were to afflict the miners' trade union movement, Stötzel also rejected the republicanism and atheism of the SPD, arguing that social reform could only be achieved on the basis of Christian thought.[72] Stötzel and his supporters were then propagating a general working-class identity aimed at appealing to the numerous factory workers of Essen as well as the miners. On the other hand, Schröder was portrayed as representing only the miners. Furthermore, by indicating that he had been selected in Bochum for the Essen seat, Schröder's credibility as a representative of the locale was undermined. At the same time claims that the Centre was somehow not nationalist were refuted, especially since the new Kaiser's social reform programme was seen as already part of the Centre's own policies.[73] This robust identity which included aspects of class, nationality and religion coupled with the overwhelmingly Catholic complexion of Essen ensured that Stötzel was comfortably returned for the constituency for the sixth consecutive time.

Mabon too played upon his labour credentials. In 1885 he defeated another Liberal and coalowner to take the Rhondda seat. At subsequent elections he repeatedly emphasised the fact that he represented the 'rights of labour'.[74] During his 1892 election campaign a leaflet specifically referring to Mabon as a 'Labour member' was distributed. Yet, despite this emphasis on labour, the class collaborative tone of Liberalism was also strongly evident. A guest speaker from Cardiff, Sir Edward J. Read, deplored the fact that Mabon was being challenged. Read's language mixed both the nationalist and labour sentiments that became characteristic of Lib-Labism in south Wales.

> They were Welshmen, he presumed. Where were they going to get a Welshman who could stand up more nobly for Wales than that candidate had done? They were labouring men – where could they pick, from one end of the land to the other, a man more worthily representative of the labouring classes than was Mabon? They were miners – where could they find a man anywhere who had done the real work which Mabon had done in Parliament and on Commissions on behalf of the miners of that country?[75]

Mabon was presented as a capable representative of both Wales as a nation and the working class. It is noticeable here that class was not being used with pejorative connotations as in Thomas's election song. As early as 1885 Mabon had claimed that the miners wanted 'class representation'.[76] Mabon had begun to introduce a new element to the language of the people used by the Liberals; one which recognised the importance of labour. Welshman and worker, Mabon represented a compound of class identity with the other political ideals of Liberalism.[77] The Liberal Unionist challenger eventually withdrew from the contest. Indeed, so secure seemed Mabon's position in the Rhondda that at the next election thought was given to whether it was necessary for him to campaign at home, when his influence might be better used to help in other contests.[78]

Yet there were tensions within Mabon's Lib-Labism. At a meeting of the Cambrian Miners' Association in 1892 he supported the issue of a parliamentary fund, complaining that Liberals who contributed to his electoral expenses expected 'a meeting for every pound sent'.[79] This desire for some degree of independence extended to policy. Where the interests of labour and Liberalism seemed to clash, the former often took precedence. Mabon, for example, had little to do with Cymru Fydd, believing that the type of Home Rule supported by this movement would damage the labour cause.[80]

## Political discourses, 1900–1914

During the January 1910 election campaign in south Wales an article appeared in the *Merthyr Express* that praised politicians such as Mabon, William Brace and Tom Richards, while castigating Keir Hardie. It claimed there was no difference in nature between Liberalism and Labour. 'Liberals are friendly to the general programme of Labour, and nothing is more certain than that this programme will never be forwarded or realised without the sympathetic co-operation of a Liberal Government and the Liberal party.' Mabon, Brace and Richards were commendable because they understood this. 'They are men who frankly recognise how much Labour owes to Liberalism and how much it has yet to gain from Liberal co-operation.' Hardie, on the other hand, completely failed to understand this. The article drew a distinction between Labour and Socialism and accused Hardie of trying to turn the former into the latter.[81]

But was this hard and fast distinction between Hardie and Mabon entirely justified? A look at Hardie's discourse suggests that he too owed much to Liberal traditions. During the December 1910 election, for example, Hardie's election address played on old Liberal shibboleths such as Irish Home Rule, Home Rule for Wales, disestablishment, temperance and educational reform. Tariff reform was rejected outright.[82] Hardie's support for that centrepiece of Liberal discourse, disestablishment, was unequivocal. During the January 1910 campaign he stated at Aberdare 'that he had been advised that it would be a wise policy on his part to say as little as possible about Disestablishment. He has been during the past session working in Parliament for the recognition of Wales as a distinct nation.'[83] Just as in Liberal discourse, Hardie sought to combine the issues of religion and nationality to create an identity that would mobilise the workers.

However, the issue of religion could be problematic for Hardie. His opponents, specifically Pritchard Morgan, the incumbent that Hardie had unseated in 1900, sought to use the perceived atheistic nature of socialism to de-legitimise Hardie's representation of the Merthyr constituency. A handbill produced by Morgan quoted thirteen reasons not to vote for Hardie. Of these four dealt with socialism's threat to religion.[84]

Morgan's attempt to portray Hardie as un-Christian was largely unsuccessful. In an open letter to Merthyr ministers in 1911 Hardie stated that 'socialism is not only not antagonistic to Christianity, but is the embodiment of everything which Christ taught concerning the life that now is'.[85] This put succinctly Hardie's view of the relationship between socialism and Christianity and represented a constant theme in all his discourse.[86] Perhaps more important than this was the support he received from a small group of socialist, or at least socialist-friendly, ministers.[87] The Revd J. Nicholas, for example, claimed that 'Christianity needs the Labour Party and the Labour Party needs Christianity'.[88] Meanwhile, the Revd George Neighbour openly regretted that the ILP faced religious prejudice and hoped through his efforts to ease that.[89] Articles linking Christianity, socialism and Welsh nationalism also appeared in Welsh language journals,[90] although generally there was a paucity of pro-socialist Welsh literature.[91] It has been rightly pointed out that the majority of Nonconformist ministers were generally hostile to socialist ideas.[92] The miners' agent Vernon Hartshorn blamed his defeat in the 1910 elections for Mid-Glamorgan on the intervention of the local ministers who had claimed 'every vote given to [him] was a vote for Atheism'.[93] A particularly vocal opponent, although a one-time ILP supporter, was W. F. Phillips, who contributed articles against socialism to Nonconformist journals.[94] However, when compared to the attitude of the Protestant and Catholic Churches towards socialism in the Ruhr, the whole stance of Nonconformity towards socialism, or at least trade unionism, seems more ambiguous. The fact that some ministers were supportive of socialist policies is of significance as it illustrates how the more fluid nature of civil society in south Wales meant that associations were able to accommodate a plurality of viewpoints. Protestant and Catholic ministers in the Ruhr did not produce articles entitled 'Why I am a Socialist',[95] and socialist ministers were a comparative rarity in Germany. Where they did engage with social issues their intervention was seen as a means of preventing the working class from being seduced by social democracy. The less than monolithic antipathy to socialism of the Nonconformist ministers, their failure to produce any effective counter-measures and the open support of a minority provided socialist ideas in Wales with some legitimacy and allowed them to be largely elided with older themes by the Welsh labour movement.[96]

Hardie thus successfully 'blended Radical and Labour appeals'.[97] But while Hardie undoubtedly adopted several of the Liberal themes, he was also careful to maintain an independent identity. Although he supported Liberal initiatives, such as the budget in 1909, he was careful to emphasise that cooperation was 'not to be taken as an indication of that [the Labour Party] had merged their independence or identity in the Liberal party'.[98] In his personal life too he tried to maintain a certain distance from the Liberals.[99] Where his discourse differed from that of the Liberals was in its emphasis upon the working class. As Peter Stead has argued, by maintaining his independence, Hardie was able 'to capture the very word "labour"'.[100] Yet, the way in which he combined this class with a nationalist identity is evident in this speech he gave at Dowlais.

> The Nationalist Party that I have in mind is this; the people of Wales fighting to recover the possession of the land of Wales, the working class of Wales acquiring possession of the mines, of the furnaces, and of the railways, of the great public works generally, and working these as comrades, not for the benefit of shareholders, but for the good of every man, woman and child within your borders. That is the kind of Nationalism I want to see brought about. And when that comes the red dragon [will be] emblazoned on the red flag of Socialism, the international emblem of the world.[101]

This first part, with its emphasis on 'the people' and 'the land', had strong resonance with the *gwerin* image supported by the Liberals. However, the rest of the speech was in sharp opposition. As already noted, much of Liberal discourse in the 1890s had little to say about class. The rural *gwerin* were placed in juxtaposition to the landowners, while the working class was subsumed within the wider community, their interests seen as coterminous with those of the employers. Hardie was effectively denying this identity of interest between employers and workers, while replacing the *gwerin* with the Welsh working class. Admittedly, elements of the communitarian ethos of Liberalism remained. Commenting on the Aberdare dispute he deplored the effects that it had upon the valley communities and claimed 'an industrial Crisis like the one now prevailing in the Aberdare Valley and elsewhere, shows how dependent the entire community is upon the working class'. The working class was now being characterised as the most important group within society rather than a part within the interdependent whole. Furthermore, the 'other' within Hardie's discourse included not only the landowners but also the capitalists. Hardie optimistically prophesied that if 'all will combine and Vote Solid for Labour, the day will soon come when there shall be neither Landlord or Capitalist left to grind the faces of the poor'.[102]

This consistent 'othering' of the employers marked the fundamental difference in the discursive practice of Hardie and his Liberal counterparts. However, during the 1900s, Liberal discourse became increasingly concerned with labour issues. The *Rhondda Leader* noted in 1903 that there appeared to be a change in political currents with the title 'Lab-Lib' increasingly replacing 'Lib-Lab', and Mabon called upon Liberal representatives to become representatives of Labour first.[103] Alfred Thomas's electoral poem for 1910, in contrast to 1895, made no mention of religion or nationalism. Instead it was focused firmly on the economic benefits that the workers could expect under free trade.[104]

For whom will you vote?
Is it a Liberal or a Tory
That will lead John Bull to glory?
This simple question is answered at once.
Since the Tory is such a dunce.

If the Tories get in first,
Then our Parliament will be the worse;
They will put tax on bread,
And hit the workman on the head.

If the Liberals win the day,
Then we'll certainly have fair play;
They'll help the poor and tax the rich,
They'll pull the workman from the ditch.

Sir Alfred Thomas, Brace and Mabon
Are not the men for the Lords to tread on;
Mabon fights on behalf of Labour,
Despite the strenuous tones of Balfour.

If we speak of old age pensions,
Tories say, 'We had the same intentions',
They promised them long years ago,
The Liberals had to give them though.

Sir Alfred Thomas is the man,
If he doesn't triumph then who can?
Vote for him and Labour too,
And you'll be paid, and we paid too.

Something of a shift is therefore identifiable within Liberal discourse. More emphasis was now being placed upon the industrial working class. It is clear then that the Liberals could, up to a point, adapt the language of class to their own needs. However, unlike Hardie, they were unwilling to accept the labour–capital dichotomy as a basis of their discourse. Older notions of the people remained strong and the conflict between the Lords and the Commons in 1910 provided an excellent opportunity for the Liberals to exploit the discursive dichotomy they had created in the nineteenth century. At the massive demonstration on 18 September 1909 at Victoria Park, Swansea in support of the 'People's Budget' the common theme was 'the People vs. the Lords'. Alfred Mond claimed that the landlords wished 'to make the people subservient to their charity and their doles, while the Liberals wanted to make them free and independent men and women, not relying on charity and doles, but on their rights as citizens'. Mond also emphasised the communitarian ideals of Liberals by the juxtaposition of the industrialist and the landlord. While the industrialist created wealth and gave employment to hundreds, the landlord created nothing. The Liberals 'wanted the burden shifted from industry to indolence, taken off the backs of the workers and put on the shirkers'.[105]

That Mond should stress the importance of the industrialist to the well-being of the community is not surprising since he was a prominent employer himself. However, Mabon, who, since the affiliation of the MFGB to the Labour Party in January 1909, had officially been a Labour MP, also used the same rhetoric.

> [The Lords] wished at all cost to reserve their power to plunder the people. Every Labour representative and every Liberal and Radical citizen in the realm must be of the opinion that the feudal age should in fact and practice come to an end. Our present system of land ownership had devastated the countryside; it had imposed heavy burdens upon industries, and cramped the development of towns and villages – simply crippled capital and impoverished labour. The House of Lords must either be mended or ended, the question being, 'Who is to govern, the Peers or the People?'

Yet Mabon prioritised his Labour identity and it was to this that he coupled notions of patriotism. He claimed that 'he was prepared to place before them the views of the people which he hoped to represent from a Labour standpoint, and he would be prepared to add everything that appertained to Welsh nationalism and every progressive measure, wherever it might come from'.[106]

Mabon's stress on Labour, however, did not go far enough for some. During the second 1910 election Mabon was severely criticised by the Mid-Rhondda Socialist Propaganda Committee. Their speakers, Ben Tillett and George Edbury, cast doubts on Mabon's Labour credentials, pointing to his lack of support for the Cambrian strikers. The election itself was 'fake' since the Liberals could be relied upon to support the capitalists. Working-class interests were both independent from and in opposition to those of the Liberals. However, ultimately the Committee did not run its own 'working-class' candidate due to insufficient funds.[107]

By the end of the 1900s a political discourse predicated on a specifically working-class identity and with some affinities to SPD language in the Ruhr was beginning to emerge during general elections in south Wales. As Chris Williams has shown in his study of the Rhondda, this discourse existed at the local level from an earlier date,[108] and influenced the campaign to affiliate the MFGB to the Labour Party. Workers elsewhere were also attempting to place members of their own class on local authorities. In Ammanford, for example, meetings of workmen decided to field ten candidates, among them five colliers, for the Ammanford Urban District Council.[109] But it was by the late 1900s that this discourse based on the working class began to make a wider impact at the national level in Wales. Two reasons can be identified as contributing towards this growth. First, the industrial disputes of 1910 and 1911 placed pressure on the Liberals' *gwerin* discourse. The strikes and the introduction of troops undermined a discourse based on cooperation and community. The Liberals, however, remained wedded to their old discourse based on class cooperation and were unable to accept one predicated on conflict between labour and capital. Instead, they sought to paper over the cracks. The pomp and ceremony surrounding the Prince of Wales's investiture in 1911 was, for example, a deliberate attempt to create a new identity that encompassed both elements of the *gwerin* and the Anglican Church.[110] Second, during the 1900s there was a growth in Labour and Socialist organisations. Although these may have disagreed over the means, they shared a common commitment to advancing the interests of the working class.[111] They were also willing to tie the social directly to the political, as Hardie did with the Aberdare strike. They used strikes and unrest to reinforce their demands for independent working-class representation. However, the discourse used by Labour MPs in this period still owed much to Liberal traditions. The *gwerin* image was not wholly abandoned. Many Labour MPs, after all, had been Lib-Labs. Rather than a clean break between Liberal and Labour discursive practice, there was a more fluid shifting of emphasis. Political organisations and trade unions were placing increasing stress on a specific working-class identity. This impacted on politicians' discourse, contributing to a growing emphasis on 'labour'. Consequently, Labour was in a better position to seize on the industrial unrest of the late 1900s and tie the social more explicitly to the political. However, rather than abandoning Liberalism's communitarian ethos, Labour was able to realign it. The community was being made synonymous with the working class, while employers replaced the landowners as the 'other'. This realignment did not reach fruition until after the war, but was under way before 1914.

The fluidity of discourse evident in south Wales, at least between Liberalism and Labour, was in marked contrast to the comparative rigidity of political languages in the Ruhr. The split between the German labour and liberal movements had occurred much earlier and by the 1890s the SPD had established its own distinctive political identity. Unlike the British Labour Party in the 1900s it did not draw much from other political currents within the region. In the years before the war it continued to predicate an identity largely based upon the workers and class conflict. It began to notch up some successes with Otto Hue elected for Bochum and Theodor Bömelburg for Dortmund in 1903 and 1907. However, during the 1900s the revisionist wing of the party did begin to utilise a more patriotic language.[112] Hue, for example, was keen to stress the reformist nature of the SPD: '[The Party] is the real reform party, although the radical reform party, in our parliament.'[113] He complained that during election campaigns the opposition accused the SPD of seeking to overthrow the state when in actual fact thousands of members and activists were diligently working for reform of the system rather than its destruction.[114] In a later article he defended Bebel's claim that, should Germany be threatened, Social Democrats would take up arms in defence of the country. Hue maintained that 'one cannot stress enough: patriotism (*Nationalbewusstsein*) – not nationalism – is not anti-social democratic'.[115] Despite this patriotic element the language of class remained central. Reformist action was meant to benefit the working class and overcome the twin obstacles of conservatism and clericalism.

By the 1900s the Polish community had also began to make an independent impact in electoral contests. The network of Polish associations now included electoral committees and in 1903 Polish candidates stood for the Essen, Bochum and Dortmund seats. In no seat did the Polish vote amount to more than seven per cent.[116] However, the Polish vote in Dortmund-Hörde was not about winning elections, but had everything to do with expressing a certain identity. As Stanley Suval has pointed out, voting was a means of affirming their separateness. 'The Polish candidate did not live in the district; he did not even have to campaign. Polish votes were designed to show the other social groupings that the Poles had arrived and were to be reckoned with.'[117] The party's discourse was consequently predicated on a specific ethnic identity.

Despite the emergence of this Polish party there was little change in the discourses of the older parties. Bömelburg, a leading figure in the bricklayers' union, used the same discursively created dichotomy in his 1907 campaign as Tölcke had used in 1893. The National Liberals were once again castigated as the 'party of the great, industrialist, German bourgeoisie'.[118] The Centre, on the other hand, was once more accused of splitting the workers.

This was despite the fact that the Centre had actually dropped their usual candidate, Lambert Lensing, for one with stronger working-class credentials. Since 1890 the party had been squeezed out of the run-off elections by the National Liberals and the SPD. This coincided with an increasing professionalisation of the Christian trade unions which led them to shake off non-worker direction.[119] As a result Johann Effert, a prominent figure in the Christian trade union movement and leader of the Commission of Seven during the 1905 strike, was selected as candidate. However, while Effert claimed to be an *Arbeiterkandidat* (workers' candidate), the Centre Party itself continued to depict itself as a *Volkspartei*. Effert was both 'a man of the people, a man of severe work'. He would represent both the workers and middle classes. In fact, the very emphasis of the SPD upon the working class was seen as a weakness within Centre discourse. 'Social Democracy, the patented 'workers' party', wants to follow a 'pure workers' politics', they show no interest for the remaining estates (*Stände*).' Effert agreed that the working class outnumbered the others in society, but argued that if they tried to pursue a purely working-class type of politics the other estates would come together to defend themselves. The workers would inevitably fail 'because capital is not on our side'. Finally, the usual stick of Christianity was used to beat the SPD. 'Because Christianity and Socialism stand in opposition like fire and water, therefore a practising Christian cannot be a Social Democrat.'[120]

Other Centre candidates, such as workers' secretary Johannes Giesberts in Essen, also stressed their working-class origins.[121] However, the discourse used in Effert's candidature demonstrates that, although the Centre Party was becoming increasingly aware of the potential advantage to be gained from selecting working-class candidates, the rhetoric used by the party remained fundamentally unchanged. Emphasis was still placed on the *Volk* and the community, of which the workers were just one, albeit a sizeable, part. The workers' interests could only be furthered through cooperation with others, an explicit rejection of the class conflict that lay at the heart of SPD discourse.

The SPD, on the other hand, rejected the idea that Effert would be able to do anything constructive for the workers in the Reichstag precisely because his party had a middle-class wing. 'First the Centre lets working-class deputies (*Arbeiterabgeordneten*) be elected, then they will neutralise them. The Centre was a middle-class party and had now become a capitalist party.' The Centre's social and economic policy was regarded as no different from that of the National Liberals. An example was made of their support for tariffs, which directly affected workers through higher prices. 'If the Catholic workers want their workers' interests represented in the Reichstag, they must attach themselves to the workers' party, Social Democracy.'[122]

Pamphlets and leaflets also depicted this working-class SPD/non-working-class Centre dichotomy. Although few of the thousands of leaflets that were printed for elections have survived, one illustrates well the discursive practice of the SPD towards the Centre. Entitled 'Who should we believe?' it attacked the Centre's claims that it was the most socially active party in Germany. Accusations of splitting the labour movement mainly revolved around the issue of Christian trade unions.

> The Centre founded the Christian trade unions. If these really sought to fulfil their purpose, i.e. to represent workers' interests, then they affect them like a spanner in the works (*Knüppel zwischen den Beine*). Where is there recorded a case of the Centre taking the side of the workers in a wage dispute? Nowhere! It is neutral – in favour of the employer. The Christian trade union leaders will be described – when it's in the interests of capitalism – exactly like the free trade unionists as liars, troublemakers and unscrupulous agitators, who bring down misfortune and despair on the workers.

The interests of the workers were placed in opposition to those of the counts, lawyers and prelates within the party. 'The demands of the workers' interest simply mean a damaging of their own interests.' Worse, however, were the Christian trade union leaders who actually betrayed the interests of their own class.[123] The Centre was not only in opposition to the working class, it actually undermined them from within.

Class was placed above religious identity. The pamphlet criticised the Centre's continual claims that religion was in danger, arguing that this was simply a cover for the interests of the capitalists and the landowners. 'Christian duty for the workers, Christian rights from the Lords. Why are the Catholic workers forbidden to unite with their class comrades to improve their material conditions? Because that would endanger religion!'[124] The SPD, on the other hand, had raised the expectations of the workers. It 'ennobled the worker; it raised him to the proper position as the bearer of the prosperity of the nation'.[125]

Interestingly, there was an echo of the religious justification used by the British Labour movement, when it was claimed that the SPD represented the 'true spirit of Christianity'.[126] However, this theme was as not extensively developed as it was in Hardie's political language and the impression is of something tagged on rather than central to Social Democratic discourse. The view of religion the SPD continually reiterated was that it was a private matter,[127] something that largely limited its use in constructing a collective identity.

SPD discourse in the twenty-two years between 1890 and 1912 appears not to have changed much. Admittedly there were attempts to reach out to other groups, such as white-collar workers. During the 1907 run-offs in Essen, for example, an appeal was made to make every effort to win both 'workers and the small officials'.[128] In 1903 SPD activist Peter Leimpeter, while criticising the other parties for abandoning the workers as soon as the elections were over, claimed that the representation of the SPD as 'un-national' was unjust. But ultimately he referred back to the workers, stating that such accusations were unjustified 'because only [the Social Democrats] care for the well-being of their voters'.[129] Just as religion remained the centre point of Centre discourses, so class remained the central pillar of SPD rhetoric.

## Summary

The striking difference between the political discourses of the Ruhr and south Wales was the comparatively 'fixed' nature of the former compared with the fluidity of the latter. Crucially, political languages in the Ruhr were largely 'milieu-fixated' while in south Wales they were 'community-fixated'. In the former coalfield, discourse was concerned with mobilising a certain constituency or subculture which consistently provided it with a core of support. Although the parties claimed to represent the community in its entirety, they invariably tried to mobilise their followers by placing one group in juxtaposition to another. Indeed, attempts to anchor their discourses in a broader community were made difficult by the rapid expansion and ethnic diversification of the Ruhr workforce before 1914. Instead the parties fell back on their most reliable supporters, an attitude encouraged by the Wilhelmine electoral system. In this manner, Catholics were contrasted to Protestants, Social Democrats to Christians, and so on. In south Wales political parties were encouraged by the first-past-the-post electoral system to appeal to a wider electorate. Parties sought to encompass the whole of the community with their rhetoric. The juxtaposition of identities in south Wales was between those perceived as belonging to the community (in Liberal discourse the workers and industrialists) and those who were not (the landowners).

The Liberal Party was able to dominate the political field through its successful linkage of nationalist and religious concerns. This found a certain resonance in the predominantly Nonconformist mining valleys. Although, when compared with the activities of Catholic priests in Germany, Nonconformist ministers may not have been so intimately involved in electioneering, their seemingly ubiquitous presence on the platforms of Liberal candidates was symptomatic of the links between Liberalism and Nonconformity.

Within Liberal political language the workers did not really emerge as a distinctive group, but were subsumed within the broader idea of the community. It was believed that what was good for the community would inevitably benefit the workers. The *gwerin* image itself was mainly rural, but it seems that its nationalist and religious components were enough to tie the majority of the miners to Liberalism until the 1900s.

Liberal discourse also showed a certain degree of elasticity in accommodating the growing language of labour. The emergence of the Lib-Lab MPs can be regarded as strengthening the appeal of Liberalism generally by placing a greater emphasis on the needs of the working class. Indeed, the Lib-Lab MPs introduced a shift in the connotations associated with the term 'class'. In earlier Liberal language it was used pejoratively and contrasted to 'the people' or 'the masses'. Mabon, however, used the concept positively, especially when he spoke of the 'labouring classes'. This positive usage became more widespread from 1900. But to claim that Lib-Lab discourse proved a stepping stone to the political language of Labour is too simplistic. There was a great deal of continuity between Keir Hardie's discourse and traditional Liberal rhetoric. What was occurring in south Wales was a re-alignment of emphasis. Gradually notions of the community and the working class were being elided, becoming synonymous. At the same time the industrialists, who had been considered part of the community by Liberalism, were being excluded and aligned with the landowners. While Liberal ideas continued to have force, labour issues were becoming more and more prominent. This was not a completely linear process, but nevertheless there seems a general tendency towards Labour. The fact that Labour did not represent a complete break with south Wales's Liberal past and actually inherited much of its progressive mantle was to lay the basis for its successes in inter-war Wales.

In the Ruhr, on the other hand, there was less interchange between political discourses. Each party had largely established its discursive identity by the 1890s, if not earlier. Both the Centre and National Liberal parties shared certain affinities with the Liberals in south Wales in the way that they combined nationalism with religion and attempted a cross-class appeal. Both parties hoped through this mixture to represent themselves as a *Volkspartei*, although in reality they were delimited by religion. However, depending on the local political circumstances, this could be either inclusive or exclusive. In Dortmund, for example, the National Liberals sought to develop a general religious identity to counter the growth of Social Democracy. In Bochum in the 1890s, however, the National Liberals used the *Volk* in a more exclusive manner, seeking to 'other' the Catholics through their perceived divided loyalties. Ultimately, although both parties considered themselves representative of the German nation, one was closely linked to Protestantism, while the other saw its explicit goal as the protection of the Catholic Church. This, coupled with the organisational structures that they had built, meant that they had a strong head start over the SPD when that party finally emerged from the wilderness of the Anti-Socialist Law.

Once able to campaign freely the SPD constructed its own identity explicitly around the working class. While the Centre and National Liberals characterised themselves as representatives of the whole community, the SPD portrayed the workers as the community and in inevitable opposition to the other classes owing to the workings of capitalism. Unlike the other parties in the Ruhr and the Labour movement in south Wales, it had no real religious element within its appeal, something that undermined attempts to encompass all the workers.

Over time there were slight changes and modifications made to these identities. As the example of Effert shows, the Centre Party became aware of the advantages of more identifiably working-class candidates in this overwhelmingly industrial region. However, Effert's candidature remained firmly bound to the traditional discursive practices of the Centre. There was little interchange with either the SPD or the National Liberals' political language. Changes in political discourse in the Ruhr occurred within parties rather than between them, as was the case in south Wales.

The need to adapt political languages was to become more pressing following the Great War. The conflict posed challenges for all the political parties as the parameters within which they operated changed. They all had to come to terms with the unprecedented circumstances of total war and the social, political and economic upheaval that followed.

## Notes

1 Cited in N. Evans and K. Sullivan, 'Yn llawn o dân Cymreig (full of Welsh fire): The language of politics in Wales 1880–1914', in G. H. Jenkins (ed.), *The Welsh Language and its Social Domains 1801–1911* (Cardiff: University of Wales Press, 2000), p. 569. The *South Wales Daily News*, the main liberal daily in south Wales, was commenting on the landslide victory for the Liberals in the 1906 general election.

2 Cited in W. Buschak, '"Vergessen wir keinen Augenblick, daß wir in Bochum sind." Gewerkschaften und Sozialdemokratie vor 1914', in Friedemann and Seebold (eds), *Struktureller Wandel*, p. 130.

3 See M. Cowling, *1867: Disraeli, Gladstone and Revolution. The passing of the Second Reform Bill* (Cambridge: Cambridge University Press, 1967), pp. 288–310; M. L. Anderson, *Practicing Democracy: Elections and Political Culture in Imperial Germany* (Princeton: Princeton University Press, 2000), pp. 4–8.

4 Anderson, *Practicing Democracy*, p. 9.

5 Hoffman, 'Politik in der Provinz', p. 230.

6 Anderson, *Practicing Democracy*, p. 226.

7 D. A. Hamer, *Liberal Parties in the Age of Gladstone and Rosebery: A Study in Leadership and Policy* (Oxford: Clarendon Press, 1972), pp. 173–4.

8 K. O. Morgan, 'Gladstone, Wales and the new radicalism', in Peter J. Jagger (ed.), *Gladstone* (London and Rio Grande: Hambledon Press, 1998), pp. 132–3.

9 G. A. Ritter, 'Die Sozialdemokratie im Deutschen Kaiserreich in Sozialgeschichtlicher Perspektive', *HZ*, 249 (1989), 296.

10 *RL*, 8 August 1903.

11 In Germany Lassalles' General Union of German Workers was formed in 1863; August Bebel and Wilhelm Liebknecht's German Socialist Workers' Party in 1868/9. In 1875 these merged to form the German Social Democratic Party (SPD). In Britain the Social Democratic Federation (SDF) and Socialist League were founded

in 1884, the ILP in 1893, and the Labour Representation Committee in 1900. For the ideological and organisational reasons behind the German workers' early break from liberalism see J. Breuilly, *Labour and Liberalism in Nineteenth-Centry Europe: Essays in Comparative History* (Manchester: Manchester University Press, 1992), p. 115, Thomas Nipperdey, *Deutsche Geschichte 1866–1918. Arbeitswelt and Bürgergeist* (Munich: Beck, 1998), pp. 352–3; W. L. Guttsman, *The German Social Democratic Party, 1875–1933: From Ghetto to Government* (London: Allen & Unwin, 1984), p. 28; and Ritter, 'Sozialdemokratie im Deutschen Kaiserreich', p. 296.

12 See Chapter 1.

13 *RWAZ*, 11 April 1893.

14 *Ibid.*, 13 April 1893.

15 Initially Schorlemer-Alst and Lensing had both refused the seat, possibly due to doubts over their chances of success, *T*, 4 April 1893.

16 *RWAZ*, 13 April 1893.

17 *RWAZ*, 3 May 1893, cited in A. Herzig, *Carl Wilhelm Tölckes Presseberichte. Zur Entwicklung der duetschen Sozialdemokratie 1848–1893* (Munich: Dokumentation, 1976).

18 Ritter, *Arbeiterbewegung, Parteien und Parlamentarismus*, p. 32.

19 For the Lassallean and Marxist trends within the early social-democratic movement see Thomas Welskopp, *Das Banner der Brüderlichkeit. Die deutsche Sozialdemokratie vom Vormärz bis zum Sozialistengesetz* (Bonn: Dietz, 2000), pp. 668–705.

20 G. A. Ritter, *Die Arbeiterbewegung im wilhelminischen Reich. Die Sozialdemokratische Partei und die freien Gewerkschaften, 1890–1900* (Berlin-Dahlem: Colloqium Verlag, 1963), pp. 128–34.

21 *RWAZ*, 7 June 1898.

22 *RWAZ*, 25 June 1898. Hilbck was a member of the virulently anti-Catholic Evangelical Union: S. Suval, *Electoral Politics in Wilhelmine Germany* (London: Univesity of North Carolina Press, 1985), p. 217.

23 See D. Hopkin, 'Labour's Roots in Wales, 1800–1900', in D. Tanner *et al.*, *The Labour Party in Wales*, pp. 54–7; John Parry, 'Trade Unionists and Early Socialism in South Wales, 1890–1908', *Llafur*, 4:3 (1985), pp. 43–54.

24 Hardie's election at Merthyr will be dealt with later.

25 *PC*, 20 November 1885.

26 It was sung to the tune of 'Auld Lang Syne'. *GFP*, 20 July 1895.

27 Thomas tabled National Institutions (Wales) Bills in 1891/2 which proposed the devolution of several administrative functions to Wales, such as a Secretary of State, national council, Welsh education department, local government board and National Museum. See J. Graham Jones, 'Alfred Thomas's National Institutions (Wales) Bills of 1891–92', *WHR*, 15:2 (1990), 218–39.

28 Evans and Sullivan, 'Dân Cymreig', p. 569.

29 Morgan, *Wales in British Politics*, pp. 120–65.

30 G. A. Williams, *When Was Wales? A History of the Welsh* (Harmondsworth: Penguin, 1985), p. 237.

31 Evans and Sullivan, 'Dân Cymreig', pp. 564–8.

32 *GFP*, 13 July 1895.

33 *PO*, 29 September 1900.

34 *MMSWA*, 9 April 1880.

35 *SWDS*, 11 July 1895. The question was in response to Morgan's claim that Welsh members had themselves defeated a recent Disestablishment Bill by adding too

many amendments. A Welsh member was considered as someone who supported specific Welsh interests and was, therefore, most commonly applied to Liberal MPs from the Principality.

36 *SWDT*, 27 September 1900.

37 *Ibid.*, 28 September 1900.

38 See *CJ*, 5 and 12 January 1906.

39 *Ibid.*, 14 January 1910.

40 *W*, 19 July 1895.

41 *Ibid.*, 7 January 1910.

42 *GFP*, 20 July 1895. 'Trech gwlad nag arglwydd' roughly translated means, 'The country beats the lord'.

43 Joyce, *Visions of the People*, pp. 62–3.

44 *Ibid.*, pp. 55–6.

45 On the politics of personality see Williams, *Democratic Rhondda*, pp. 50–6.

46 *ME*,17 March 1888.

47 *RWZ*, 28 January 1890.

48 The bill, aimed at increasing the strength of the army, sprung from military concerns following the lapsing of the Reinsurance Treaty with Russia. See G. A. Craig, *Germany, 1866–1945* (Oxford: Oxford University Press, 1978), pp. 257–8.

49 *MS*, 5 June 1893.

50 *RWZ*, 6 June 1893.

51 *DZ*, 2 May 1893.

52 *EVZ*, 3 February 1890.

53 'An die rheinische Handwerker', printed in *ibid.*, 31 May 1893.

54 Clipping, STAD 9046.

55 Suval, *Electoral Politics*, pp. 217–18.

56 *T*, 4 June 1898

57 *Ibid.*, 18 June 1898.

58 *Ibid.*, 23 June 1898.

59 *DZ*, 9 June 1898.

60 *Ibid.*, 14 June 1898.

61 *Ibid.*, 21 June 1898.

62 *RWZ*, 9 June 1893.

63 *RWZ*, 15 June 1893. Krupp himself did not actively engage in the election campaign.

64 'Der Zweck heiligt die Mittel', STAB, I 1 b/29.

65 'Theure evangelische Glaubengenossen', STAB, I 1 b/27.

66 *MS*, 3 June 1898.

67 *Ibid.*, 13 June 1898.

68 'Bergleute, Kameraden', STAB, I 1 b/26.

69 'Arbeiter! Kameraden', STAB, I 1 b/28.

70 See Williams, *Democratic Rhondda*, pp. 32–6; Jäger, *Bergarbeitermilieus und Parteien*, pp. 68–75.

71 *EVZ*, 3 February 1890.

72 *Ibid.*, 4 February 1890.

73 *EVZ*, 18 February 1890.

74 Williams, *Democratic Rhondda*, pp. 32–45.

75 *GFP*, 9 July 1895.

76 *PC*, 6 November 1885.

77 Chris Williams, 'Democracy and nationalism in Wales: The Lib-Lab enigma', in R.

Stradling, S. Newton and D. Bates (eds), *Conflict and Coexistence. Nationalism and Democracy in Modern Europe: Essays in honour of Harry Hearder* (Cardiff: University of Wales Press, 1997), p. 111.

78 *GFP*, 20 July 1895.

79 *GFP*, 2 July 1892.

80 See Williams, 'Democracy and nationalism', pp. 118–25; C. Williams, *Democratic Rhondda*, p. 45.

81 *ME*, 15 January 1910.

82 Election Address, 3 December 1910, SWCC SC 491.

83 *AL*, 8 January 1910.

84 Handbill, GRO D/D Vau 25/10a.

85 *MP*, 13 May 1911.

86 For the religious motifs within Hardie's discourse see R. Pope, *Building Jerusalem*, pp. 8–14.

87 See handbill addressed to the SWMF in which it was claimed 'Mr Keir Hardie has the support of leading Ministers of the Gospel of all Denominations', GRO D/D Vau 24/15 and the handbill entitled 'Church Socialist League Manifesto', GRO D/D Vau 24/13.

88 *RL*, 18 May 1907.

89 *Ibid.*, 14 September 1907.

90 See for example 'Sosialaeth Gristionogol', *Y Geninen*, April 1908, vol. 26, pp. 107–112 and 'Y Ddraig Goch a'r Faner Goch: Cenedlaetholdeb a Sosialaeth', *Ibid.*, October 1911, vol. 29, pp. 10–12.

91 Evans and Sullivan, 'Dân Cymreig', p. 581.

92 Pope, *Building Jerusalem*, p. 59.

93 P. Stead, 'The Language of Edwardian Politics', in Smith (ed.), *People and a Proletariat*, p. 156.

94 See 'Y Ddraig Goch ynte'r Faner Goch? Sosialaeth a Chenedlaetholdeb', *Y Geninen*, October, 1911, vol. 29, pp. 254–8 and 'Y Perygl oddiwrth Sosialaeth yng Nghymru', *ibid.*, January 1912, vol. 30, pp. 7–9.

95 'Why I am a Socialist' by Reverend G. Jones, *RS*, 7 December 1912.

96 On this conflation of religion and socialism see C. B. Turner, 'Conflicts of faith? Religion and Labour in Wales 1890–1914', in Hopkin and Kealey (eds), *Class, Community and the Labour Movement*, pp. 67–85.

97 D. Howell, *British Workers and the Independent Labour Party, 1888–1906* (Manchester: Manchester University Press, 1983), p. 253.

98 *NGMGH*, 2 October 1909.

99 Hardie wrote to D. A. Thomas in response to an invitation to dine with leading Liberals in the Merthyr constituency – 'My position in the Labour movement is such that I feel the need for great caution and so deeply am I convinced of the need of keeping the movement clear of any taint of Party politics that I have for years past made it a rule not to accept such invitations', enclosed with letter from D. A. Thomas to Mr Rice, 10 April 1902, GRO, D/D Xes 5/2.

100 Stead, 'Language of Edwardian politics', p. 152.

101 *MP*, 14 October 1911.

102 Election Address, 3 December 1910, SWMA SC 491.

103 *RL*, 19 September 1903.

104 *GFP*, 28 January 1910.

105 *NGMGH*, 25 September 1909.

106 *RL*, 8 January 1910.
107 *RL*, 10 December 1910.
108 Williams, *Democratic Rhondda*, pp. 72–81.
109 *CJ*, 20 March 1903.
110 J. S. Ellis, 'The Prince and the Dragon: Welsh National Identity and the 1911 Investiture of the Prince of Wales', *WHR*, 18:2 (1996), 272–94.
111 Williams, *Democratic Rhondda*, p. 68.
112 Lützenkirchen, *Der sozialdemokratische Verein für den Reichstagswahlkkreis Dortmund-Hörde*, pp. 45–6.
113 *Sozialistische Monatshefte*, September 1906, p. 725.
114 *Ibid.*, p. 726.
115 *Ibid.*, September 1907, p. 699.
116 H.-W. Tiedtke, 'Die Reichstagswahlen und die Reichstagwahlkämpfe im Südlichen Teil des Ruhrgebietes von der Gründung des Nord-Deutschen Bundes bis zum Beginn des Ersten Weltkriegs' (Hausarbeit der Fachprüfung für das Lehramt am Gymnasium, Bochum, 1976).
117 Suval, *Electoral Politics*, p. 213.
118 *RWAZ*, 3 January 1907.
119 Schäfer, 'Milieu der katholischen Arbeiter', pp. 215–18.
120 *T*, 2 January 1907.
121 See *EVZ*, 7 and 23 January 1907.
122 *RWAZ*, 10 January 1907.
123 '"Wenn sollen wir glauben?" Eine Flugschrift für das Volk in Stadt und Land von einen Arbeiter', IZfD, M11. The pamphlet is undated but was edited by Max König, SPD *Bezirkssekretär* for Dortmund-Hörde, and printed in Dortmund sometime after 1907.
124 *Ibid.*, pp. 7–8.
125 *Ibid.*, p. 11.
126 *Ibid.*, p. 13.
127 *Ibid.*, p. 12.
128 *Arbeiterzeitung*, 1 February 1907.
129 *Amt* Weitmar to *Landrat* Bochum, 8 May 1903, STAB, LA 1323.

# 4

# The Great War and the post-war years, 1914–1926

The threat of the trenches and the drafting to the army created a psychological state, which was everywhere favourable to the Spartacist agitation.[1]

*Reichskommissar* Carl Severing on how the bad experiences of the miners during the war fed the extreme left.

The circumstances of the war intensified the men's convictions that they were being exploited by their employers and by the Government. They believed that they were morally and legally entitled to a share in the boom of prosperity which, as they alleged, the war had brought to their employers.[2]

Revd J. Vyrnwy Morgan commenting on the Welsh miners' determination to use their bargaining position to enforce higher wages.

## Introduction

The Great War had a profound impact on the lifeworld of the miners and civil society in the two regions. The conflict accentuated the process of fragmentation and homogenisation that had emerged over the previous two decades. The strength of the SWMF allowed the miners to win substantial concessions from the government and the employers. On the other hand, the stronger position of the Ruhr industrialists, coupled with the divisions in the German miners' movement, meant the unions in the Ruhr were unable to take advantage of their better bargaining position to force substantial concessions.

The war also reshaped the political arena. It not only gave fresh impetus to organisation for the labour movement in south Wales, but it also provided an opportunity for activists to demonstrate their competence and credibility. In Germany, however, support for the war divided the SPD. This split laid the foundations for the emergence of the Communist Party, which challenged the SPD's claim to represent the workers. Underpinning these political changes in the Ruhr was the fragmentation of the social milieux that had characterised Wilhelmine Germany. The experience of strikes, lockouts, unemployment and political violence produced newer, but no less deep divisions among the Ruhr miners. On the other hand, the generally positive experience of the war in south Wales consolidated support for the Labour Party's case for nationalisation and strengthened the appeal of the SWMF. Moreover, the common experience of unemployment and industrial action generally helped reinforce the more cohesive social structures and strengthen support for Labour.

### The Great War and the miners

Since the impact of the war on both coalfields has been explored in detail elsewhere, this section will limit itself to drawing several general points of comparison.[3] Key among them is the assumption that, despite the loss suffered by individual families, the miners of south Wales as a whole improved their economic and political position during the war. On the other hand, the various miners' unions in the Ruhr were unable to capitalise on their industry's important contribution to the war effort. Rather than consolidating their position as the SWMF did in south Wales, they became more fragmented.

In both countries the war was portrayed as one of national defence. Yet despite these calls to unity the miners' movements reacted differently to calls to abandon pre-existing disputes. The Ruhr unions' acceptance of the *Burgfrieden* (literally Peace of the Fortress) was symbolised by the unions' decision to give up their right to strike.[4] In Britain, on the other hand, the MFGB refused to give up its right to strike, while the SWMF even stood aside from a guarantee to seek arbitration in disputes.

The different attitudes of the British and German unions should be seen in the context of their status prior to the war and their bargaining position during it. The British unions were well established within society and recognised by employers and state alike. Moreover, a steady supply of coal was crucial to maintaining the British fleet's blockade of Germany, thereby placing the British unions, and especially the SWMF, in a better position to demand concessions. Many of the German unions, on the other hand, were still struggling to achieve recognition on the eve of the war.[5] The low level of union density in the Ruhr also placed the mining unions in a weaker position vis-à-vis the state and government. Put simply, the Ruhr unions had fewer opportunities to turn the situation to their advantage. They hoped that full cooperation with the state would win them recognition and were therefore willing to give up more than their British counterparts.[6] The *Bergarbeiterzeitung* estimated in December 1914 that 30,000 of its members were at the front and claimed that after the war they could never again be regarded as enemies of the Reich.[7] This hope was shared by the SPD. Hue, who toured the region explaining the need for continued production to the miners, exemplified this attitude. At Bochum he justified SPD support for the war because 'before the war the Social Democrats were accused of being unpatriotic fellows and revolutionaries. After the war [they] will not tolerate such suggestions and will demand equality with other parties'.[8] At a meeting in Altenbochum in 1917 it was claimed that the unions 'must behave perfectly. After the war the Alter Verband members will hopefully not be treated as third and fourth-class citizens anymore.'[9] Similarly, *Gewerkverein* activist Hermann Vogelsang's pamphlet, *Der Weltkrieg: Deutschlands Bergarbeiter im Kriege* (The World War: Germany's Miners at War), dwelt on the sacrifices the miners were making for the Fatherland both at the front and at home.[10] Indeed, it is difficult to see what other options apart from wholehearted cooperation were available to the *Alter Verband* and *Gewerkverein* in 1914. Failure to cooperate would have played into the hands of employers, National Liberals and other opponents of unionism, who could have cited such a refusal as proof of the un-German nature of both the unions, the SPD and the Centre Party.

We need not be detained by the details of the British and German governments' intervention in the industry.[11] Generally the British Government proved more willing to intervene in the coal industry and when it did it was usually to the benefit of the miners. Following the failure of the Coal Mining Organisation Committee (CMOC) to ensure harmonious relations within the industry, the government intervened in July 1915 to end a strike in south Wales over wages. With further unrest brewing the government established control of the south Wales coalfield at the end of 1916. Government control did not entail substantial changes in the way the industry was run,[12] but miners gained materially as their wage demands were largely accommodated. One recalled his high pay due to the war wages: 'I was earning one pound two and nine say when I was fourteen, and when I was fifteen I was earning four pounds two and nine with the awards.'[13] These increases eventually put the British miners' wages above the rate of inflation and increased their real earnings.[14]

In the Ruhr, employers continued to reject recognition, despite the unions' participation in the *Burgfrieden*. In 1915 a proposal for a joint-conciliation board was rejected by employers.[15] Government officials thought it a 'highly dangerous experiment and Von Sydow, Minister of Commerce, fearing that it would disrupt production, ordered the issue dropped'.[16] Wage demands were similarly rejected, or countered with claims that production must first be increased.[17] Such rejections led the *Bergarbeiterzeitung* to question why there was no mutual understanding in the mining industry. An editorial demanded why employers were unwilling to meet the unions. 'We have to ask this after more than three years of war! Everywhere else the will to accommodation is present. In the mining industry the old spirit of lordship (*Herrengeist*) reigns as before!'[18]

The employers and unions agreed on why this was so. In 1915 the *Alter Verband* pointed to the MFGB successful wage demand and claimed that this depended upon the strength of that organisation. If the German miners were as organised as the English then their position would have been very different.[19] Similarly, the employers were claiming as late as 1918 'that the number of organised workers is too small for them to be recognised as the representatives of the entire workforce',[20] a view supported by the authorities. The unions complained that while the employers took that view, they also made it as difficult as possible to organise the workers.[21] The owners also continued to play on political fears by characterising the British unions as solely economic and the

German unions as social-democratic, despite the SPD support for war credits.[22]

The trade unions in the Ruhr were further undermined by the importation of foreign and POW labour and the use of threat of the draft. By the end of the war some 70,000 POWs were employed in the mines.[23] At some pits forced labour made up almost half the workforce.[24] Industrialists also colluded with the military authorities to have troublesome employees removed. For example, three union representatives on the workers' committee of the Welheim mine were drafted soon after their election.[25]

Some recognition of the trade unions' role in the war effort came in 1916 with the introduction of the Auxiliary Service Law. Part of the Hindenburg Programme to increase production, it incorporated some important concessions, such as the exclusion of company unionists from works' committees.[26] Initially the unions' reaction was positive,[27] but by April 1917 the *Bergarbeiterzeitung* felt obliged to publish an article defending the law from Independent German Social Democratic Party (*Unabhängige Sozialdemokratische Partei Deutschlands*, USPD) accusations that it straitjacketed the working class.[28] In fact, the law proved a double-edged sword for it limited the miners' ability to move between mines in search of better conditions.[29]

As a consequence of these developments, the position of the Welsh miners compared with their counterparts in the Ruhr was comparatively good. Not only did British miners' real earnings rise,[30] they also suffered less from food shortages. The effect of shortages upon German morale has been questioned, but it seems clear that the Welsh miners suffered less privation than the Ruhr miners did. The wildcat strikes in the Ruhr from 1916 primarily concerned issues of food distribution.[31] It was only towards the end of the war that the strikes developed political overtones.[32] The unions claimed that the strikes were the work of unorganised miners,[33] but other evidence suggests that many were organised and points to a loss of control by the unions over their members.[34] By contrast social disorder over food in south Wales was more limited and became apparent only from the end of 1917.[35]

Furthermore, the issue of food supply and distribution provided an issue upon which the labour movement in south Wales could coalesce. The Food Control Committees (FCCs) established to oversee rationing and price control were initially dominated by Liberal and Tory local notables, such as shopkeepers, precisely the individuals who were alleged to be profiteering.[36] These sentiments developed into demands for increased working-class representation upon the FCCs. Workers in Tredegar, for example, passed a resolution requesting the resignation of the local FCC due to its 'unrepresentative composition from the standpoint of the working class'.[37] As pressure grew for independent working-class representation, Trades and Labour Councils (TLCs) developed a new role as guardians of working-class interest.[38] Returning to Tredegar, the local TLC soon moved its attention from increased labour representation on the FCCs to other local political bodies.[39]

Meanwhile, in the Ruhr, the labour movement was becoming increasingly fragmented as the SPD split over continued support for the war. Opposition became increasingly vocal. One small anti-war group was centred on Karl Minster in Duisburg.[40] Minster had an outlet for his anti-war sentiments in the *Niederrheinische Volkstimme*. However, Minster was removed from the editorship in early 1916.[41] This pattern was repeated elsewhere and ultimately the operation of the press bureau for the Rhineland and Westphalia was suspended on 1 July 1916 because there lay 'inside the social-democratic movement an ever sharper political antagonism'.[42]

The importance of the control of the party organs is highlighted by the example of August Erdmann, deputy for Dortmund. In 1917 he joined the USPD and consequently became a non-entity in the Majority SPD press. This tactic seems to have been effective as the local Dortmund organisation was not deeply affected by the schism and the USPD remained weak.[43] In other areas the USPD was more successful. In Gelsenkirchen it held conferences on 18 May and 12 November 1916, the latter prompting Hue to resign as SPD candidate for the Bochum-Gelsenkirchen constituency. On 6 May 1918 a USPD association for the electoral district of Bochum-Gelsenkirchen-Hattingen-Witten was established.[44]

In south Wales, anti-war sentiment began to increase from 1917. A conference of the British Council of Soldiers and Workers Delegates was held in Swansea in July 1917, but was disrupted by pro-war supporters.[45] The anti-war movement was spearheaded by up-and-coming activists such as James Griffiths at Ammanford, S. O. Davies, checkweigher at the Great Mountain Colliery,[46] and A. J. Cook.[47] However, unlike the Ruhr, these tensions in south Wales were largely contained within the existing labour movement. None of these leaders proposed the establishment of a separate political party.

The Great War ultimately had a divisive effect upon the mining communities of the Ruhr, while serving to strengthen labour organisations in south Wales. The war did see some advances for the Ruhr miners' unions. The four unions drew closer together,[48] as demonstrated by their common front against the company unions in the safetymen elections of 1915.[49] The Auxiliary Service Law also granted the unions a degree of recognition, leading to a subsequent increase in trade union membership in the later years of the war.

However, in comparison with the miners' movement in south Wales, the Ruhr miners had not been able to force significant concessions from the employers or government during the conflict. First, the individual miner was financially worse off. Second, miners' contact with the state had generally been unfavourable. It usually occurred in the context of food shortages and the draft. Third, the recognition granted to the unions by the employers was grudging. Fourth, the mining workforce itself experienced a process of restructuring through the introduction of female, youth, foreign and POW labour. As a result the Ruhr miners' experience of the war was one of 'decline in real wages, hunger and increasing numbers of accidents'.[50] These factors, coupled with the union's decision to give up the right to strike, undermined their ability to present themselves as the protectors of the miner's interests. As a result, alternative forms of action, such as wildcat strikes and workforce meetings, began to increase, as evidenced by the strike waves of 1917 and 1918. These strikes built upon older forms of interest articulation,[51] and provided fertile ground for syndicalists and communists. Finally, although the Centre Party emerged from the war largely unscathed, those miners represented by the SPD were split between the Majority German Social Democratic Party (MSPD) and USPD.

The Welsh miners, by contrast, came out of the war with a sense of confidence. Significant victories had been won in terms of wages and organisation. Unofficial strikes did not pose an organisational threat to the SWMF and should be seen in terms of prodding the union leadership towards a more aggressive policy. The positive experience of state control fed a determination to win nationalisation of the industry in the post-war years. Finally, although there were tensions between pro- and anti-war supporters in the labour movement, these lacked the intensity of comparable divisions in the Ruhr and did not take organisational form. Just as before the war, potentially divisive discourses were contained within the existing structure of the labour movement, which was strengthened by the activities of members on TLCs and other local bodies. Their activities raised the profile of the Labour Party, while participation in the FCCs revealed the division of interest between the Liberal 'shopocracy' and the working class.

The pressures of war served to accentuate the processes of fragmentation and homogenisation outlined in the previous chapters. Consequently, the lifeworld of the Ruhr became even less synchronous as working conditions deteriorated. Although the *Alter Verband* and the *Gewerkverein* moved closer together they found it increasingly difficult to control the wildcat strikes during the second half of the conflict. Furthermore, fragmentation occurred along new lines. Both the SPD and the *Alter Verband*, for example, now faced a challenge from the left in the form of the USPD and syndicalists. In south Wales, on the other hand, the lifeworld of the Welsh miners appears to have become more synchronous. Their improved bargaining position consolidated the SWMF. Incorporation of the craft unions further strengthened the union. In the political sphere, Labour organisations gained credibility through their work in the community. This sowed the seeds for the replacement of the Liberal dominance with a Labour hegemony, although this was not to come to full fruition until the 1920s.

## The post-war years, 1919–1926

The dramatic events of the first half of the 1920s in south Wales and the Ruhr have been extensively covered. The very different histories of the two regions during this period make narrative comparison difficult. Rather than providing a detailed account of these events, this section will concentrate on some of the main features of post-war industrial relations and politics and their effect upon the lifeworld of the miners and civil society.

## Industrial relations and trade unionism

Although the German mining unions had been unable to win full recognition during the war, the revolution changed the context of negotiations. The unions informed employers that without concessions they would be unable to control their members. Following tough negotiations, an eight-hour day including winding time, increased bonuses for overtime and a minimum wage were agreed on 14 November 1918. On 23 November 1918 the Stinnes-Legien agreement effectively recognised the trade unions.[52] Wage negotiations would now take place within the context of the *Zentralarbeitsgemeinschaft* (Central Works Community, ZAG).

In south Wales the industry remained under state control until 1921. The positive experience of government control was clear in the miners' demand for nationalisation. In fact, the desirability of nationalisation had been raised in 1912. In 1918 the SWMF revived the idea at the MFGB conference in Southport.[53] Nationalisation became a key policy goal of the MFGB, and the Sankey Commission gave the impression that its hopes might be realised. However, by 1919 the government was freed from its dependence on Welsh coal. Consequently, a tendency to side with the employers was emerging as it sought to return to the pre-war status quo.[54]

Whatever the attitude of the state, the SWMF had in nationalisation a rallying call for its members. Industrial conflict could be largely simplified to a struggle between the union and the employers. The trade unions in the Ruhr, on the other hand, lacked this focus. The ZAG provided the *Gewerkverein* and the *Alter Verband* with the recognition that they had craved and on that basis they felt obliged to protect it. But the Ruhr unions faced pressure both from employers, who sought to win back the concessions granted during the revolution, and from their own membership as the pattern of localised strikes continued after the war. Demands included increased wages, a six-hour shift, socialisation and the establishment of mine councils. The Thyssen mines in Hamborn were particularly prominent in this respect and their strikes in late 1918 and 1919 had a domino effect upon neighbouring pits.[55]

Furthermore, although the company unions had largely been eliminated by the agreement, the *Gewerkverein* and *Alter Verband* faced a new challenge from syndicalist-communist unions, which capitalised on the workers' discontent with the established organisations' policies. Two main oppositional unions can be identified, although the boundaries between them were fluid. The FVdG re-emerged after the war to play an important role in the 1919 strikes in Hamborn.[56] It later merged with several other syndicalist unions in September 1919 to form the *Freie Arbeiter-Union Deutschlands* (Free Workers' Union Germany, FAUD).[57] By the end of 1919 the FAUD had some 30,000 Ruhr miners among its membership.[58] By 1923 it had grown to 57,127 members with 143 branches in the Ruhr. But metalworkers now formed the majority, comprising 59.8 per cent of total membership, while miners made up 33.9 per cent. Furthermore, while the metalworkers had 52 branches with an average size of 657, the miners had 70 with an average membership of 277, suggesting that the latter were more thinly spread.[59] The years 1919 and 1920 represented a peak for syndicalism in the Ruhr,[60] while in south Wales its anti-state doctrines, if not its fighting spirit, were largely abandoned in favour of state collectivism.[61]

The second radical union was the *Union der Hand- und Kopfarbeiter* (Union of Hand and Head Workers). Tensions between communist and syndicalist elements within the FAUD led to a split in 1920 and the formation of *Freie Arbeiter Union Gelsenkirchen* (Free Workers' Union Gelsenkirchen, FAUG). The FAUG was more inclined towards communism than syndicalism and in September 1921 it merged with two other unions to form the communist *Union*.[62] It appears that the majority of miners previously organised in the FAUD made the move to the FAUG and later the *Union*. By the end of 1920 the *Union* organised approximately 66,000 miners in the Ruhr.[63] In the mining sector, most of the communists were organised in the *Union*, although it proved an unruly body.[64]

Miners' unionism in the Ruhr in the early 1920s was therefore just as prone to fragmentation as it had been during the *Kaiserreich*. The syndicalist-communist organisations posed a real challenge to the established unions' control of works councils and the *Knappschaft*. The *Union* was able to win almost 35 per cent of works council seats in 1924 (see Tables 4.1, 4.2 and 4.3). These results disguise the extent of the leftist union's challenge in individual mines. The Dortmund police authorities reported in 1924 that the Free Trade Unions had 'lost more than 40 per cent of their earlier votes. Especially evident is the decline of the mining association'. In Dortmund the *Alter Verband* came second to the syndicalists in all but one mine. The combined effects of disappointment with the ZAG, the emergence of syndicalist-communist unions and the demographic fluctuations in the workforce can be clearly seen in the declining membership of the three main pre-war unions (see Table 4.4).

After an initial surge in support towards the end of the war, the pre-war trade unions saw a steady decline in membership after 1919/1920.

Unfortunately, figures for the *Gewerkverein* are fragmentary, but it appears that it proved better at retaining membership than the *Alter Verband* or the ZZP. This impression is reinforced by analysis of the Knappschaft and works' council

*Table 4.1* Results of elections to the *Knappschaft*, 1921 and 1924

| | 1921 | 1924 |
|---|---|---|
| Votes | 205,959 | 213,330 |
| *Alter Verband* | 96,806 (47%) | 93,082 (43.6%) |
| *Gewerkverein* | 61,264 (29.7%) | 61, 918 (29.0%) |
| ZZP | 8,766 (4.3%) | 3,195 (1.5%) |
| *Hirsch-Duncker* | 2,213 (1.1%) | 2,795 (1.3%) |
| Left Radicals | 36,910 (17.9%) | 52,340 (24.5%) |

*Source:* Jäger, *Bergarbeitermilieus*, p. 47.

*Table 4.2* Percentage of seats won on works' councils in the Ruhr mining industry by the trade unions, 1921–1926

| | 1921 | 1922 | 1923 | 1924 | 1925 | 1926 |
|---|---|---|---|---|---|---|
| Percentage participation | – | – | – | 71.9 | 75.6 | 76.0 |
| Free trade unions (including *Alter Verband*) | 45.7 | 41.2 | 41.7 | 32.2 | 40.5 | 66.5 |
| Christian trade unions (including *Gewerkverein*) | 18.9 | 18.0 | 20.6 | 21.2 | 23.2 | 25.5 |
| *Hirsch-Duncker* | 0.7 | 1.1 | 1.3 | 1.6 | 1.3 | 1.6 |
| ZZP | 7.5 | 5.9 | 3.0 | 1.2 | 0.7 | 0.7 |
| *Union* | 27.0 | 26.0 | 32.9 | 34.3 | 29.1 | – |
| Syndicalists | – | 4.7 | – | 7.3 | 3.0 | 2.9 |
| Company | 0.1 | 0.4 | – | 0.4 | 0.5 | 1.7 |

*Source:* Jäger, *Bergarbeitermilieus*, p. 49.

elections. The *Alter Verband* and the ZZP, on the other hand, recorded a decrease in membership and voters. The syndicalist-communist nature of the FAUD and *Union* suggests that they had greater success in attracting members from the social-democratic *Alter Verband* than the religious *Gewerkverein*. Although caution must be exercised in drawing too direct a link between *Alter Verband* losses and FAUD/*Union* growth, there is some evidence of movement from one to the other. Former *Alter Verband* official Johann Spaniol, for example, became leader of the *Föderation der Bergarbeiter Deutschlands Syndikalisten* (Federation of German Miners, Syndicalist). Its membership in the Oberhausen area was almost twice that of the *Alter Verband*.[65] The government agents reported that communist agitation within the members' meetings of the *Alter Verband* and other unions had reached such proportions that the associations felt obliged to warn their members against such actions.[66]

A KPD member in Hamborn hinted at the depth of rivalry between the old unions and the syndicalist-communist organisations arising from the revolution. 'When he is in the pit, he always notices the relentless enmity between the SPD and the KPD. Cooperation with the SPD is impossible. Our dead comrades must shame us if we went along with the SPD.'[67] Organisational rivalry sometimes shaded into physical confrontation as at the Westerholt mine where syndicalists forced their way into the works' council room and threatened the *Alter Verband* and *Gewerkverein* representatives. Such issues became tests of strength between the unions.[68] During the February 1919 strike the *Alter Verband* felt compelled to organise security units to protect its members. A miner was injured at an incident at the *Alter Verband* building in Bochum between these units and armed miners.[69]

The divisive nature of trade unionism in the Ruhr during this period is highlighted by their different conceptions of socialisation. Initially, there appeared to be unanimity. On 10 January 1919 the local SPD, USPD and communists spontaneously formed a nine-man commission in Essen to prepare for sociali-

*Table 4.3* Results of works' council elections to some Dortmund mines, 1924

| | Free union | | Christian unions | | Syndicalist | | *Union* | | Not Organised | | *Hirsch Duncker* | |
|---|---|---|---|---|---|---|---|---|---|---|---|---|
| | Vote | S | Vote | S | Vote | S | Vote | S | Vote | S | Vote | S |
| Mines | | | | | | | | | | | | |
| Minister Stein | 462 | 2 | 277* | 1 | 945 | 5 | 317 | 2 | 190 | – | 71 | – |
| Dorstfeld I/IV | 460 | 5 | 246 | 2 | 515 | 5 | 418 | 4 | | | | |
| Dorstfeld II/III | 395 | 3 | 130 | 1 | | | 505 | 5 | | | | |
| Tremonia | 460 | 3 | 147 | – | 1,201 | 8 | | | | | | |
| Hansa | 451 | 1 | 450 | 2 | 740 | 6 | | | | | | |
| Hardenberg | 638 | 5 | 207 | 1 | 1,300 | 7 | | | | | | |
| Glückauf | 333 | 2 | 220 | 1 | 1,017 | 8 | | | | | | |
| Kaiserstuhl I/II | 423 | 2 | 133 | 1 | 1,384 | 8 | | | | | | |
| Hohenzollern | 255 | 2 | 233 | 2 | 633 | 5 | | | | | | |
| Hansemann | 357 | 5 | | | 1,113 | 8 | | | | | | |
| Kaiser Friederich | 720 | 3 | | | 920 | 7 | | | | | | |

Key: S = seats

* At Minister Stein the *Gewerkverein* and the ZZP candidates stood on a joint ticket.

Source: *Polizeiverwaltung* to *Regierungspräsident* Arnsberg, 28 April 1924, STAM OBA 14481.

*Table 4.4* Membership of the *Alter Verband, Gewerkverein*, ZZP and SWMF, 1919–1926

| Year | *Alter Verband* | *Gewerkverein* | ZZP | SWMF |
|---|---|---|---|---|
| 1919 | 159,136 | | 46,261 | 185,044 |
| 1920 | 163,356 | 73,859 | 45,000 | 197,668 |
| 1921 | 152,255 | 73,420 | 45,018 | 117,610 |
| 1922 | 107,174 | | 33,200 | 87,080 |
| 1923 | 97,251 | | | 147,611 |
| 1924 | 67,305 | | 5,000* | 148,496 |
| 1925 | 65,585 | | 3,500* | 129,155 |
| 1926 | 60,821 | | 3,000* | 136,250 |

*Approximate figures.

*Source:* Jäger, *Bergarbeitermilieus*, p. 339; Williams, *Capitalism*, pp. 88–9.

sation, while a conference of Workers' and Soldiers' Council representatives, union and Government representatives unanimously adopted a resolution supporting it three days later.[70]

Despite superficial consensus, the diverse understandings of socialisation were clear to both union leaders and outside observers. Imbusch commented at the time that the word had become a slogan (*Schlagwort*) that hid many different meanings. For some it meant the abolition of private property, for others workers' control, for others nationalisation.[71] Bergassessor von Lowenstein commented the following year that 'the slogan of socialisation possesses too great an attraction, [it] has become too indispensable a means of recruitment (*Werbemittel*) for the socialist party and the free trade unions to drop it in their struggle for existence against the radical movements of Unionists and Communists'.[72]

The government initially refused to recognise the commission, but did establish its own Socialisation Commissars. The miners accepted the commissars in a second conference on 20 January and the government eventually recognised the commission on 14 February. In the meantime, spontaneous attempts to occupy some mines caused tensions between the commission members.[73] But it was events in Hervest-Dorsten that precipitated the split between the SPD and left-wing representatives. The murder of the president of the colliery clerks' association, Kohlmann, provided the army with the excuse to occupy the area. The resulting violence and death of forty miners following the arrival of the *Freikorps Lichtschlag*, was condemned by a conference in Mülheim on 16 February and a general strike proclaimed. At the miners' conference two days later the USPD and KPD delegates supported the proclamation, while the SPD representatives attempted to pass a resolution condemning it. The refusal to allow a vote on the resolution led the SPD members to resign from the Commission effectively ending the cooperation.[74]

The SWMF was spared this organisational fragmentation, although there were dissenting voices within the organisation.[75] Government sources reported 'support of nationalisation is not unanimous. In south Wales and Lancashire there is a good deal of opposition.'[76] Unrest was further attributed to younger miners. Government sources noted that resistance to nationalisation in south Wales actually came from two directions: syndicalist and conservative. The former rejected nationalisation in favour of 'mines for the miners'.[77] However, the authorities concluded that 'examples of opposition to nationalisation, however, must at present be regarded as a district exception to the general miners' attitude'.[78]

It is also significant that policy differences did not lead to the establishment of separate organisations. In fact, in 1919 and 1920 the SWMF was successful in merging some of the smaller craft unions into its structure.[79] The SWMF ambition to absorb the smaller unions was succinctly put by J. D. Morgan at the 1919 May Day demonstration at Ystradgynlais. 'The days of the small Union were numbered. He believed that the whole of the men in the coal mining industry should come into the Miners' Federation – whether they worked underground or on the surface.'[80] Consequently, membership continued to grow in this period.

The defeat suffered by the miners in the lockout of 1921, however, began to reverse the process of amalgamations that had occurred since the war and fostered a lack of interest in unionism among some quarters.[81] Moreover, the Enginemen, Stokers and Craftmens' Association was resurrected after the strike as the South Wales and Monmouthshire Mechanical and Surface Workers' Union, before reverting to its original name in 1924. The Mechanical Union's leader D. B. Jones not only claimed that his members sought more cooperative, peaceful relations with employers,[82] but also played upon occupational differences. The union's General Secretary claimed the miners had 'no conception of their work. The Federation catered for those employed at the coal face, forgetting that a strong arm and a thick head could cut coal, whilst it required brains to drive an engine.'[83]

Although employers did not make the re-establishment of the craft union easy, the SWMF took its formation seriously and conducted a vigorous, but largely unsuccessful campaign to destroy it.[84] The 1926 strike saw a further fragmentation of the SWMF following the formation of the South Wales Miners' Industrial Union. The following years were characterised by a struggle both against this 'scab' union and non-unionism in an effort to rebuild the SWMF.

## Work processes and miners' communities

The inter-war period saw changes in the work process in both regions. Production dipped in the Ruhr in the post-war years following the release of the forced labourers, a regression in mechanisation and the fact that most of the easily accessible seams had been worked out during the war.[85] Employers sought to maintain and increase production by expanding the workforce. The number of underground workers increased from 261,348 in 1918 to 384,833 by 1921, surpassing its pre-war level. Meanwhile, the surface workforce swelled from 90,770 in 1918 to 141,388 by 1922.[86]

From the mid-1920s the Ruhr coal companies increasingly sought to tackle rising costs through rationalisation and mechanisation. In the two year between 1922 and 1924 the workforce fell by 80,000.[87] The *Gelsenkirchener Bergwerks AG* provides a good example of the spread of mechanisation in the industry. Although the distribution of mechanisation was uneven, by 1925 52.9 per cent of its total production was machine cut, 24 per cent was won by shot-firing, while the remaining 23.1 per cent was won by hand. Of the machine-won coal the overwhelming majority (91.9 per cent) was cut using a pneumatic pick.[88] Mechanisation undoubtedly made a strong impact, both in terms of coal winning and hauling, on the Ruhr coal industry, but did not totally eliminate handworking. Alexander Stenbock-Fenmor, a student, worked briefly in the Ruhr mines. His description of mining in the early 1920s clearly illustrates the continued existence of the *Kameradschaft* style of working.[89]

The south Wales workforce also expanded in 1919 and 1920. The Housing Committee of the Ogmore and Garw UDC, for example, anticipated a total increase of 7,810 at the collieries in the area in 1918/19.[90] In terms of mechanisation, however, south Wales lagged behind the Ruhr. The first cutting equipment was seemingly only introduced during the war.[91] However, if geological conditions meant the Welsh owners were slow to introduce cutting technology they were quick to utilise advances in conveyors. The introduction of conveyors also meant that the hewers' small-group style of working remained relatively untouched, thereby avoiding potential conflict.[92] Klaus Tenfelde has argued that changes in the working methods before the war led to a loss of autonomy and solidarity among the miners through deskilling. However, as Chapter 1 has shown, work processes in south Wales and the Ruhr were not dissimilar before the war. It was in the 1920s, with the widespread mechanisation of the Ruhr mines, that the experience of work really began to diverge between the two regions. While the hewers of south Wales largely continued to work in small groups, the German miners had to adapt to new techniques and machinery with all the consequences for autonomy, skill and solidarity that Tenfelde suggests. When considered in the terms of the comparison, it appears that the process of deskilling in the Ruhr is better located in the 1920s, rather than in the pre-war period.[93]

While some miners struggled to adapt to the new technology, others had to deal with the other consequence of rationalisation: unemployment. Interestingly, German historians have regarded the experience of unemployment during the Weimar Republic as divisive. The unemployed were increasingly isolated not only from the workplace but 'also from the organisational and cultural traditions of the social-democratic labor movement'. Those who were still employed enjoyed the benefits of what wage increases there were as well as social welfare schemes such as municipal housing. As a result there developed 'something of a "respectable" working-class culture distinct from the "rougher" street culture'.[94] This is in sharp contrast to Britain where historians have identified a lessening of the stigma attached to drawing relief in those areas of endemic unemployment during the inter-war years. Instead unemployment was 'experienced collectively'.[95] Unemployed miners could remain members of the SWMF and were therefore not segregated from their employed colleagues. The regionalised nature of unemployment in Britain and the more diverse industrial base in the Ruhr suggests that divisions between employed and unemployed were not as great in south Wales as in the Ruhr. The Llanbradach Workmen's Hall, for example, established a committee to devise a scheme allowing the unemployed and their families to use the facilities.[96] However, unity in unemployment should not be over-emphasised. In 1923 the same committee passed a resolution refusing free entry to the unemployed, apart for a few elderly men who had failed to find work.[97]

The unions in both areas also had to contend with demographic changes. The expansion of the Ruhr workforce in the early 1920s led to an influx of young, unmarried miners, new to the job.[98] This new workforce was more amenable to ideas of direct action than the discipline of trade union work. Tenfelde characterises the turbulent circumstances in the Ruhr during the early years of the Weimar Republic as a '*Generationskonflikt*'. Yet this movement also drew on older traditions in the mining movement through its use of workforce meetings and direct democracy.[99] Lucas, in his study of Hamborn and Remscheid, also places the appeal of direct action in the former down to the relatively youthful proletariat, whose experiences of discontinuity and insecurity in life made direct action seem the best means of immediate redress.[100] In fact, some *Bergämter* complained that, in their rush to recruit, some companies often made no attempt to distinguish between peaceful and restless workers.[101]

Yet, while the employers were trying to increase the workforce, there is also evidence suggesting that many were seeking to leave the Ruhr. In 1919 several of the *Bergämter* reported on active advertising for emigration to South America, France and elsewhere.[102] The Polish milieu was particularly affected by this emigration. The Bochum police estimated in 1922 that some 25,000 Poles had left the area.[103] Such emigration undermined the Polish milieu and provoked an existential dilemma in the 1920s.[104] The ZZP went into terminal decline in the mid-1920s as it found it difficult to maintain its membership due to the emigration of the Poles to the resurrected Polish State and industrial regions outside Germany.[105]

However, the Polish subculture did not completely break up. The depleted Polish organisations continued their efforts to maintain their Polish identity. The Polish Workers' Party, for example, continued to raise the issue of Polish language education at its meetings, although much time was taken up with expressions of support for the Polish State.[106] Furthermore, in some areas, such as Bottrop, Polish communities formed part of communist strongholds.[107] While contemporaries often regarded the colonies as an obstacle to organisation before the war, Brüggemeier argues that after the war both the SPD and KPD were able to make inroads. Some even became 'red' strongholds.[108] The pre-war Polish milieu found itself squeezed between a loss of its constituent members through emigration and the competing attractions of other socio-political subcultures.

South Wales, on the other hand, experienced a more general loss of population prompted by the downturn in the fortunes of the coal industry. Yet these demographic changes served to further homogenise the mining communities by reinforcing the Anglicisation of the region.[109] The process of secularisation prior to the Great War also continued in the 1920s. Although the denominations recognised that they had to engage in interdenominational social work, their main focus remained the education and salvation of the individual.[110] However, the effects of the decline of the industry and the large strikes of 1921 and 1926 placed the emphasis on collective rather than self-help. It is probably overstating the case to claim that the chapels lost influence simply because 'the chapel people didn't believe in anything outside it'.[111] During the strikes the chapels did perform important relief work. Ultimately, however, they were unable to convert their principles into enduring practicalities.[112]

Secularisation was a feature of both regions and was aided by the continuing spread of secular forms of entertainment. In the 1920s the number of miners' cinemas in south Wales increased despite the industrial defeats.[113] The workers' appreciation for sport remained undiminished. The football teams of the coastal cities were a focus for regional identities for the south Wales miners.[114] Sporting teams, however, did not escape the effects of deprivation. Teams like the Ammanford rugby club suffered following the 1925 and 1926 strikes as many players left the area in search of work.[115] This was despite the relative success of the 1925 strike and the fact that the anthracite coalfield was less affected by the decline in coal exports than the rest of the Welsh coalfield.[116]

Although some sports teams in the Ruhr, such as Schalke 04 were regarded as politically neutral, other teams and sporting associations in the Ruhr were linked to certain political parties.[117] Unlike Britain where the workers' love of sport was seen as 'the healthiest antidote to extremist teaching',[118] in the Weimar Republic sporting teams often formed part of the fragmented political culture. Although sports associations in the Ruhr were not simply recruitment schools for the parties, during the 1920s some associations became the foci for power struggles between communists and Social Democrats.[119] Following the dissolution of the *Union*, the communists were reported as 'gouging' social-democratic cultural and sporting organisations from within in an attempt to take them over.[120] However, in other areas such as entertainment, there was a trend towards the commercialisation of leisure and the creation of an apolitical mass culture.[121]

In south Wales the long strikes provided ample leisure time for the establishment of workers' cricket and other sports teams.[122] For some 'sport' and leisure became a means of survival. Rees Davies and his brothers, for example, worked as boxers to supplement their income.[123] In fact the miners adopted a wide variety of survival techniques. During the 1926 strike the Social and Entertainment Committee of the Garw Valley alone put on 130 free concerts, fourteen road races, seven whist drives, ten dances, two gymnastic displays, eight boxing tournaments, five drama performances and a carnival.[124] In the Ruhr there were similar efforts at self-help and fund-raising. Soup kitchens, fund-raising events, stealing from the workplace, looting of shops and the working of outcrop coal, all formed part of the miners' repertoire of self-help.[125] Yet, despite similarities in the ways the miners tried to make ends meet during the industrial disputes of the 1920s, the actual impact of those strikes upon the communities differed markedly. Francis and Smith have argued that the collective effort of the 1926 strike in particular helped create an alternative society within the coalfield.[126] How alternative this society was is debatable, but it is clear that industrial defeat led the miners to turn increasingly to Labour to satisfy their interests. Other historians have stressed how the labour movement captured positions of authority from the pre-war elite. Gilbert, therefore, sees the local miners' lodge and Labour Party as edging out the influence of the chapels and employers in Ynysybwl.[127] Similarly, the socialist Query Club in Tredegar is depicted as waging a concerted campaign to establish control of local institutions.[128] Charlie Gibbons' resolution that 'appeals from Industrial, Political and educational bodies will not be considered or entertained unless such bodies are exclusively controlled by the workers' might represent an extreme, but it does illustrate the labour movement's commitment to extending its influence independent of other groups.[129] Despite the weakening of the SWMF through the re-emergence of craft unionism, on the political level a new Labour hegemony was reinforced.

In the Ruhr, on the other hand, the strikes of 1919 and 1920 and the *Ruhrkampf* of 1920 were not a cohesive experience and led to increased political and union fragmentation. The British strikes of 1921 and 1926 occurred under Coalition and Tory governments respectively, leaving Labour's political reputation relatively untainted and allowing its representatives to play a supportive role to the miners.[130] The 1919/20 strikes and the *Ruhrkampf* of 1920, however, occurred under an SPD-led government. The bitterness aroused by the use of troops and *Freikorps* further fractured the labour movement and reinforced the formation of a communist milieu. The *Union* and the formation of communist organisations to rival those of the SPD underpinned this milieu. Admittedly, in 1924 the KPD, following Comintern instructions, ordered communists to join existing unions rather than establish radical alternatives and the subsequent break with the syndicalist-communist *Union* led to the decline of that organisation. In contrast to south Wales, where industrial events reinforced Labour's position, in the Ruhr political events would seem to have actually lessened the fractured nature of the union movement. However, this proved short-lived and in 1928 the idea of communist unions was revived in the form of the *Revolutionäre Gewerkschafts-Opposition* (Revolutionary Trade Union Opposition, RGO).[131]

## Politics

The changed electoral system in Germany and the extended franchise in Britain opened up new opportunities for the political parties in the coalfield. The Representation of the People Act of 1918 expanded the British electorate to include many more workers. There has been much debate on the implications of electoral reform for the fortunes of Labour.[132] Whichever position is adopted it is clear the parties were now chasing more votes. In Germany reform was more extensive. Adult suffrage was introduced for the over-20s, the three-class system abolished in Prussia and a proportional representation list system introduced.

An overview of the political developments in the two regions illustrates a continuation in the trends of homogenisation and fragmentation. In south Wales, the Labour Party was eventually able to build on its pre-war successes in local government and usurp the Liberal dominance, transforming the mining constituencies into a Labour stronghold. In the Ruhr, the socialist labour movement was split. Initially the SPD were successful in winning almost forty per cent of the vote, while the USPD could only manage four per cent. By the 1920 Reichstag election, however, the SPD vote had collapsed (see Table 4.5). They were overtaken by the USPD and faced a new challenge following the KPD decision to participate in elections. By the 1924 elections the USPD had dwindled to a splinter party, the majority of its moderate members rejoining the SPD, its radicals the KPD.

*Table 4.5* Results of Reichstag elections in the Ruhr in percentages, 1919–1924

| | 1919 | 1920 | 1924 | 1924 |
|---|---|---|---|---|
| Participation (%) | 86.1 | 81.8 | 79.5 | 78.5 |
| SPD | 37.2 | 14.7 | 11.3 | 17.4 |
| USPD | 4.1 | 17.7 | 1 | 0.5 |
| KPD | | 3.7 | 21 | 12.8 |
| Centre | 24.4 | 23.3 | 19.6 | 21.6 |
| BP* | 20.1 | 18.4 | 17.9 | 21.5 |
| NSDAP | | | 2.1 | 1 |

*Under BP (Bürgerliche Parteien) Jäger includes the German Nationalist Party (DNVP), the German People's Party (DVP), the German Democratic Party (DDP), the Economic Party (WP) and the Christian Socialist Peoples' Party (CSVD).

*Source:* Jäger, *Bergarbeitermilieus*, p. 30.

Labour's success in the south Wales mining constituencies was based partly on advances it made both in terms of organisation and prestige during the war. The confidence of the SWMF and the revitalised TLCs provided a sound basis for Labour to expand its political activity, although others like William Harris appealed for the formation of local Labour Parties and better organisation among women.[133] To this end the *Colliery Workers' Magazine*, established by the SWMF in 1923, carried a regular women's column. Often written by Elizabeth Andrews, the Labour Party's Women's Organiser in Wales, the column focused on themes that were regarded as having a special salience for women, such as housing and pithead baths.[134] Thus, while seeking to involve women, the magazine also revealed a gendered conception of their role, one that revolved around domestic and welfare concerns. Harris also revealed a somewhat chauvinist view of women's meetings when he complained that they often had a 'gossipy character, whilst if they meet jointly with men, this weakness does not exist'.[135]

Despite Labour's organisational advances victory was not guaranteed and varied between localities. In Ammanford, for example, all six candidates for the UDC election were defeated in 1919 and it was another two years before Labour finally controlled the municipal authority.[136] By contrast, such was Labour dominance of the neighbouring Cwmamman UDC from 1919, that the *Amman Valley Chronicle* questioned whether the local TLCs were 'not ruling the destiny of the area'.[137]

Apart from the continued strength of Liberalism in some local authorities there was also some resistance to Labour politics from within the labour movement. Die-hard syndicalists like Charlie Gibbons opposed political action as an unwelcome diversion, rued the money spent on election campaigns and tried to force the Ferndale Lodge to use its funds solely for industrial matters. That his resolution failed reflects the importance attached to political action, especially after the SWMF's first heavy post-war defeat.[138]

Local elections in the Ruhr reflected the fragmentation of the socialist movement. In the 1919 Bochum city elections the SPD formed the second largest party just behind the Centre. By 1924 they had slipped into joint third place behind the Centre and KPD, although it was able to regain a lead over the communists by 1926 (see Table 4.6).

*Table 4.6* Results of the elections to the Bochum city council 1919–1926

| | March 1919 | | May 1924 | | June 1926 | |
|---|---|---|---|---|---|---|
| Party | % votes | Seats | % votes | Seats | % votes | Seats |
| Centre | 37.6 | 25 | 27.3 | 15 | 27.7 | 18 |
| SPD | 35.4 | 24 | 13.2 | 7 | 22.4 | 14 |
| *Bürgerliche Parteien* | 22.6 | 15 | 27.4 | 13 | 22.5 | 13 |
| KPD | | | 23.7 | 13 | | 9 |
| Polish Party | 4.4 | 2 | | | 14.2 | |
| Other | | | 8.4 | 3 | 13.2 | 6 |

*Source:* Gerhard Stein, Die Wahlen zum Stadtparlament, zum Preussischen Landtag und zum Reichstag in der Stadt Bochum in den Jahren 1924 bis 1933, Bochum, July 1970, pp. 15 and 22.

At the parliamentary level Labour successfully dominated the majority of the mining constituencies from 1918. Pontypridd, Neath and Merthyr Tydfil initially fell to Coalition Liberal candidates, while Aberdare re-elected former radical Charles Stanton. Stanton, who had joined the National Democratic Party in 1916, based his appeal on patriotism and anti-Bolshevism and sought to mobilise ex-servicemen. His support proved transitory and in subsequent elections this former firebrand was forced to rely on Conservative support. This proved inadequate and he lost the 1922 election to the Labour candidate George Hall.[139] Since the miners formed a majority (or at least a substantial part) of the electorate in the constituencies shown in Table 4.7, it is reasonable to assume that a majority gave their vote to Labour.[140]

*Table 4.7* Seats contested by Labour, 1918–1924: percentage of Labour vote in south Wales mining and semi-mining constituencies

| | 1918 | 1922 | 1923 | 1924 |
|---|---|---|---|---|
| Caerphilly | 54.8 | 57.2 | 58.7 | 59 |
| Gower | 54.8 | 54.2 | 59.1 | 57.2 |
| Neath | 35.2* | 59.5 | 62.3 | Unopposed |
| Ogmore | Unopposed | 55.8 | Unopposed | Unopposed |
| Pontypridd | 42.8* | 47.2 | 54.9 | 55.9 |
| Aberdare | 21.4* | 57.2 | 58.2 | 61.6 |
| Merthyr Tydfil | 47.3* | 53 | 60.1 | 59.8 |
| Rhondda East | Unopposed | 55 | 71.9 | Unopposed |
| Rhondda West | Unopposed | 62.1 | 65.4 | Unopposed |
| Abertillery | Unopposed | Unopposed | Unopposed | Unopposed |
| Bedwellty | 53.6 | 63 | 67.6 | Unopposed |
| Ebbw Vale | Unopposed | 65.4 | 65.6 | Unopposed |
| Pontypool | 39.0 | 40.6 | 50.6 | 52.6 |

* Labour lost these elections.

*Source:* D. Tanner, 'The Pattern of Labour Politics' in Tanner, Williams and Hopkin (eds), *Labour*, p. 119.

Due to the plurality of parties in the Ruhr and the more diverse socio-economic structure of the region it is much more difficult to draw direct links between the miners and a particular party. Some contemporaries and historians have suggested that the miners were well represented among left-wing organisations.[141] Contemporary, Hans Spethmann, believed the miners to be particularly involved with the Ruhr Red Army.[142] The Army initially emerged out of the general strike called by the SPD in March 1920 to oppose the *Putsch* led by Wolfgang Kapp. It quickly developed a dynamic of its own and refused to lay down its arms once the immediate threat to the Republic had receded. Rudolf Tschirbs supports Spethmann's assertion that the miners were heavily involved in the army, arguing that the 80,000 strong force was predominantly made up of miners.[143] Others, however, have questioned the extent of the miners' involvement.[144] Of the sixteen leaders that Lucas mentions in his study of the army only two are described as miners.[145] Fragmentary evidence from courts martial following the conflict reveals that in Recklinghausen fifty out of sixty-seven prisoners were described as *Bergmann* (miner), and one as *Steiger* (mining foreman). In Essen, on the other hand, miners made up only forty-five out of 182 court-martialled prisoners, although they seemed to form the single largest occupational group.[146] The different proportion of miners, 76 per cent against 25, probably reflects the greater occupational diversity in Essen as opposed to Recklinghausen. It seems, therefore, that miners were not over-represented among the army. Moreover, participation in the army did not automatically equate with political radicalism and can be viewed as a defensive reaction to the presence of troops. The SPD-led government feared that events in the Ruhr might precipitate a civil war or provide a pretext for occupation by the Allies and ordered the *Freikorps* and *Reichswehr* into the area with tragic results.

There is some impressionistic evidence that points to wider KPD support from the miners. Stenbock-Fermor met several committed communists in the mine, but they were the exceptions to the norm. He found the miners generally apathetic and believed they harboured a strong distrust of 'intellectuals', a term which encompassed the KPD leadership. He believed that the miners supported the KPD as the lesser of other evils. He also hints at a generational difference among the workforce, for while he sees the communists as the most strongly supported group, some of the older workers retained SPD sympathies.[147] Other evidence suggests that the miners, while an important source of communist support, were not overly represented among activists. Only in three areas were communist city councillors predominantly miners (see Table 4.8). Furthermore, reports on the strength of KPD cells in the mines suggest that they were in decline by 1924.

*Table 4.8* Communist members of city councils

| City council | No. of Communist members 1924 | Miners | No. of Communist members 1926 | Miners | Total council members |
|---|---|---|---|---|---|
| Bochum | 13 | 1 | 9 | 3 | 53/60* |
| Dortmund | 20/13 | 4/2 | | | 68 |
| Duisburg | 14 | 1 | | | 63 |
| Essen | 20 | 5 | | | 78 |
| Gelsenkirchen | 20 | 5 | | | 58 |
| Herne | 12 | 3 | | | 44 |
| Mülheim | 12 | 1 | | | 51 |
| Oberhausen | 11 | 4 | | | 48 |
| Recklinghausen | 10 | 7 | 7 | 3 | 44/47* |
| Wanne-Eickel | 11 | 8 | | | 47 |
| Wattenscheid | 7 | 7 | 6 | 5 | 28/44* |

* The councils were expanded for the 1926 elections.

*Source:* Beatrix Herlemann, Die kommunalpolitik Aktivitäten der KPD im Ruhrgebiet, 1914–33, Ph.D, Ruhr-Universität Bochum, Appendix. Not all cities had elections in 1926.

Of the thirty-eight shafts in the Gelsenkirchen area only six were reported to have KPD cells. Membership in the area had fallen from 1,876 to 558, while in Buer it had declined from 1,850 in 1923 to 300 by 1924. Membership in Bochum had shrunk from 1,200 in May 1923 to 450 in 1924. The police report concluded that the decline in KPD support was due to fear of the police and French occupation authorities.[148]

The report also reveals some of the difficulties the KPD had to operate under. The party's main regional organ, the *Ruhr-Echo*, was periodically banned, while demonstrations were broken up and leaders victimised. As a result the party became increasingly isolated from the workplace and took its activities on to the streets. Moreover, the party's basis of support came increasingly from the unemployed as it made concerted efforts to organise those out of work.[149] A police report on a meeting of the Essen KPD recorded the leadership's view that 'the majority of the membership is at the moment unemployed. The KPD is actually the party of the unemployed'. Its membership stood at 7,600, but out of 29,230 miners in the Essen area only 339 were KPD members.[150]

As Table 4.5 illustrates, the decline in the SPD vote roughly equalled the increase in the USPD/KPD vote. The Centre's votes, on the other hand, remained fairly even between 1919 and 1924. Admittedly, this is a somewhat crude analysis as it does not account for first-time voters, nor does it provide a breakdown of votes by occupation. It does appear, however, that the KPD drew substantial support from the miners who had previously voted SPD. It is also clear that the SPD maintained a core of support and that by the second election in 1924 it was beginning to recover some ground (see Table 4.5).

While the KPD represented a formidable challenge to the SPD, the Communists in south Wales did not become a major challenge to Labour. Several reasons can be cited for this difference. First, divisions between pro- and anti-war figures in the south Wales labour movement did not have the same degree of acrimony as they did in the Ruhr and did not linger on into the 1920s. Morgan Jones, for example, had served eighteen months as a conscientious objector during the war, yet became Labour MP for Caerphilly in 1921 and held the seat until his death in 1939.[151] Secondly, although south Wales experienced a degree of violence during the 1920s it was not comparable to the armed conflict in the Ruhr. Violence occurred not between rival political groups, but was directed against the blacklegs who broke the norms of the labour movement. The small communist minority participated in the strikes of 1921 and 1926 alongside their Labour counterparts. The psychological gulf that emerged between the SPD and KPD as a result of political violence was avoided. Thirdly, communists in south Wales did not seek to establish counter-organisations that posed a direct threat to the SWMF and Labour Party. Arthur Horner believed that the party was never 'a breakaway from either the Trade Union Movement or the Labour Party. In the early years, communists remained members of the Labour Party'.[152] The powerful position of the SWMF immediately after the war made the notion of a separate union movement a non-starter. Instead, like the syndicalists in the pre-war movement, the communist activists sought to work through the existing labour organisations.

Despite this the extent to which the Welsh Communists remained rooted within the traditions of the pre-war labour movement can be exaggerated.

Tensions certainly existed as illustrated by Hartshorn's resignation, allegedly due to 'an atmosphere of hate and suspicion which has been created against leaders who believe in steady revolutionary reform'.[153] Following the 1918 election he was also at pains to distinguish Labour from Communist, claiming that accusations of Bolshevism were 'a disservice to the far greater number of sober Labour leaders'.[154] Tom Richards, SWMF Secretary, expressed a similar view when he criticised Communist policy, claiming 'you can destroy capitalism only by destroying the people of this country. When we have sense enough to take hold of the reins of government of this country, depend upon it there will be no need to shout class warfare or advocate revolution!'[155]

Yet, as the next chapter will show, despite these tensions the most striking aspect of the Communists' relationship with Labour was the degree of openness and engagement.

## Summary

The Great War accentuated the pre-war trends visible in political and industrial developments in south Wales and the Ruhr in the period 1919 to 1926. First, there was greater continuity in the lifeworld of the Welsh miners than in that of the Ruhr miners. Secondly, despite an observable erosion of the old milieux that characterised Wilhelmine society, the emergence of new social divisions meant that civil society in the Ruhr remained comparatively more segmented than in south Wales.

In terms of trade unionism, the strength of the SWMF coupled with the importance of Welsh steam coal for the war effort gave the union a superior bargaining position. When employers sought to resist wage increases the government interfered to the benefit of the union. Other British miners' unions also benefited from wage increases during government control as the state sought to secure supply. As a result the unions were able to expand their membership. In south Wales this included the absorption of the smaller craft unions in the coalfield. The British miners came out of the war with an overwhelming sense of confidence, which was expressed by MFGB support for nationalisation. By the end of 1922 this confidence was severely dented and the position of the MFGB began to deteriorate. In south Wales the craft unions broke away from the SWMF. The Nottinghamshire and south Wales coalfields suffered a more debilitating blow during 1926 with the formation of the Miners' Industrial Union.

However, even the defeats of the SWMF in 1921 and 1926, and the consequent unemployment and hardship, added to the synchronous nature of the lifeworld of the Welsh miners. The commonality of these experiences generally helped to draw the mining communities closer together. Moreover, there was greater continuity in the Welsh miners' experience of work. Although the introduction of conveyors affected the hauliers, the hewers' small working groups were largely untouched.

In comparison to the SWMF the mining unions in the Ruhr had less opportunity to turn circumstances to their own advantage. They were able to win some partial victories, such as the Auxiliary Service Law, and secured the longed-for recognition by the employers through the ZAG. The collapse of the *Reich*, the German Revolution and transition to parliamentary democracy initially offered new opportunities as the employers sought to bring the unions onside in order to counter more radical movements. However, these events also unleashed new challenges that undermined what gains the pre-war trade unions and parties had made. The *Alter Verband*/SPD, and to a lesser extent the *Gewerkverein*/Centre, were caught in the unenviable position of having to placate their members, while at the same time consolidating the fledging Republic. Unfortunately, these demands were often contradictory.

Both during and after the war the employers were able to use the argument that the mining unions did not represent the whole of the workforce. The structure of the mining workforce in the Ruhr also underwent substantial change. The employment of female, youth, foreign and POW labour during the war undermined trade union attempts to exert influence over the employers and the state. Since the unions were unable to protect their members' interests, and the majority of miners were not members anyway, the workers turned to short-lived strikes to force higher wages and better rations from management. The strikes, through their use of workforce meetings and direct democracy, tapped into an older tradition of miners' protest and were particularly amenable to syndicalist tactics. The flood of young workers into the area immediately after the war further strengthened the appeal of direct action. Direct action should not, however, be confused with political radicalism. In fact, it appears that the previously quiescent metalworkers of the Ruhr were actually more radical than the miners in the post-war period.

Nevertheless, while *Alter Verband* and *Gewerkverein* drew closer together, the miners' movement was split on the left by the emergence of syndicalist-communist unions. Strikes were generally divisive affairs that set one group of miners against another (or one generation against another). Finally, inter-union dialogue was undermined by the myriad, competing interpretations of socialisation. By 1924 the threat posed by these radical organisations had changed. While rationalisation and stabilisation weakened the syndicalists, the KPD tried to take their struggle against the Social Democrats into the Free Trade Unions in 1924.

On the political level, the war strengthened Labour in the mining constituencies. Through their efforts on local authority boards and on the TLCs, Labour was able to demonstrate its ability to see to the needs of the working class and support their interests. The party also benefited from the growing strength of the SWMF and the close links between the two was mutually beneficial. Both regionally and nationally, Labour also escaped an enduring schism between pro- and anti-war camps. The relationship between the Communists and Labour Party in south Wales was indicative of the comparatively fluid civil society. For most of this period double membership was tolerated, although in other British industrial areas there were greater and earlier tensions between the Communists and Labour. Although there were some tensions between Labour and the small groups of communist activists in south Wales, both groups worked together during the strikes. The community endeavour of these strikes coupled with the decline or incorporation of the older elite based on the chapel, allowed a strongly pro-Labour political culture to emerge in the mining valleys.

This toleration of multiple group membership is in sharp contrast to the Ruhr. In Germany the war caused an enduring and acrimonious national split in the SPD. The war was a much more controversial issue for the SPD than for the Labour Party and the pretence that it was a defensive conflict became increasingly hard to sustain after the defeat of Russia. The comparatively segmented nature of civil society and the leadership's desire to impose harmony on the organisation meant that the SPD found it more difficult to tolerate plural discourses within its own organisation. An example of this was the party's decision to shut down social-democratic papers which criticised the war effort. The dissidents were forced to split from the SPD if they were to air their own discourse. Although many of the USPD eventually rejoined the SPD, the radical majority filled the KPD ranks. For this group, memories of the revolution and the Government use of *Frei Korps* against communists in Berlin and the Ruhr were raw and immediate. Formal co-operation between the Communists and the SPD was not an option. The KPD sought to create, or capture organisations, to create its own milieu to rival those of the SPD and the Centre. Part of the KPD strength came from the older Polish milieu, which was increasingly dividing into nationalist and socialist camps, despite the emigration of many Polish workers. In the 1920s the KPD became increasingly a party of the unemployed and streets rather than the workforce, but the process of rationalisation in the mines provided a large potential electorate. As Weitz argues, a consequence of increasing unemployment was a social-cultural division in the German working class. In south Wales unemployment was a collective experience that served to strengthen Labour; in the Ruhr it created new divisions.

These experiences, however, had to be interpreted and reinforced by the trade unions and the political parties. The identities they created were crucial as attempts to interpret the events around them. Issues such as nationalisation, socialisation, national identity and class were the key concepts that made up the stuff of the identities they sought to create.

## Notes

1 C. Severing, *1919–1920: Im Wetter und Watterwinkel. Aufzeichnungen und Erinnerungen des Saatsministers A. D. Carl Severing, ehemaligen Reichs und Staatsminister in Befehlsbereich des VII Armeekorps* (Bielefeld: Greven, 1927), p. 12.

2 Revd J. Vyrnwy Morgan, *The War and Wales* (London: Chapman and Hall Ltd, 1916), p. 12.

3 L. S. James, 'War and industry: A study of the industrial relations of the mining regions of South Wales and the Ruhr during the Great War, 1914–1918', *LHR*, 68:2 (2003), 195–215.

4 G. D. Feldman, *Army, Industry and Labour in Germany 1914–1918* (New Jersey: Princeton University Press, 1966), p. 28.
5 Geary, 'The industrial bourgeoisie', pp. 149–51.
6 Weitz, 'Conflict in the Ruhr', p. 68
7 *BZ*, 12 December 1914.
8 *Polizeiverwaltung* Gerthe to *Landrat* Bochum, 8 November 1915, STB, LA 1323.
9 *Polizeipräsident* Bochum to *Oberbergamt*, 13 November 1917, STAM OBA 1792.
10 H. Vogelsang, *Der Weltkrieg: Deutschlands Bergarbeiter im Kriege* (Essen: Sekretariat Sozialer Studentenarbeit, 1917).
11 See A. Marwick, *The Deluge: British Society and the First World War*, 2nd edn (London: Macmillan Press Ltd., 1991); G. H. D. Cole, *Labour in the Coal-Mining Industry, 1914–1921* (Oxford: Clarendon Press, 1923); R. Redmayne, *The British Coal-Mining Industry During the War* (Oxford: Clarendon Press, 1923); Feldman, *Army, Industry and Labour*; H. Spethmann, *Zwölf Jahre Ruhrbergbau. Aus seiner Geschichte von Kriegsanfang bis zum Franzosenabmarchen 1914–1925. Bd. 1: Aufstand und Ausstand vor und nach dem Kapp-Putsch bis zur Ruhrbesetzung* (Berlin: Reimar Hobbing, 1929).
12 G. Holmes, 'The First World War and government coal control', in C. Barber and L. J. Williams (eds), *Modern South Wales: Essays in Economic History* (Cardiff: University of Wales Press, 1986), pp. 206–21.
13 J. Evans, Abercraf Community Study, SWML, AUD 221, p. 2.
14 See James, 'War and industry', p. 198.
15 *BZ*, 20 February 1915.
16 *Regierungspräsident*Arnsberg to *Oberbergamt*, 1 December 1914, STAM OBA 1814. See Feldman, *Army, Industry and Labour*, pp. 84 –5.
17 *BZ*, 10 August 1916 and 7 October 1916.
18 *Ibid.*, 11 August 1917.
19 *Ibid.*, 27 March 1915.
20 *Oberbergamt* to Minister for Trade and Industry, 23 February 1918, STAM OBA 1792.
21 Ibid.
22 *Zechenverband* to *Oberbergamt*, 10 August 1910, STAM OBA 1814.
23 *Bergbauverein* to Reichskommissar for coal distribution, 19 December 1918, STAM OBA 1160.
24 Workforce of the Wehofen shaft 1 May 1917 and *Gewerkschaft Deutscher Kaiser* 3/7 and *Rönsberghof*, June 1917, STAM OBA 1159. Here they made up 46 per cent. See K. Rawe, *'Wir werden Sie zu Arbeit bringen'. Ausländerbeschäftigung und Zwangsarbeit im Ruhrkohlenbergbau während des Ersten Weltkrieges* (Essen: Klartext, 2005).
25 *Verband des Bergarbeiter Deutschlands* to *Oberbergamt*, 17 November 1916, STAM OBA 1853.
26 See Feldman, *Army, Industry and Labour*, Ch. 4 for the full story of the laws' passage. See pp. 535–41 for the full draft of the law.
27 *BZ*, 9 December 1916.
28 *BZ*, 7 April 1917.
29 Hartewig, *Das unbrechenbare Jahrzehnt*, p. 47.
30 James, 'War and industry', pp. 195–215.
31 Hauptvorstand des Gewerkvereins christlicher Bergarbeiter Deutschlands, *Denkschrift über die Ursachen der Arbeits-Einstellungen in Ruhrkohlenbergbau*

(Essen, 1917).

32 J. Reulecke, 'Der Erste Weltkrieg und die Arbeiterbewegung im rheinisch-westfälischen Industriegebiet', in Reulecke (ed.), *Arbeiterbewegung am Rhein und Ruhr*, pp. 205–39.

33 Polizei Präsident Bochum to *Oberbergamt*, 13 November 1917, STAM OBA 1792; *Zechenverband*, 22 August 1918, BBA Bestand 20 563.

34 *Zeche Helene* und *Amalie* reported that seventy per cent of its striking workforce was organised and that the strike was led by a works' committee and *Alter Verband* member, Herzfeld. *Zeche Helene und Amalie* to *Zechenverband*, 26 August 1918, BBA Bestand 20 563.

35 The *SWDN* reported on 20 December 1917 that a large but peaceful crowd had broken shop windows by sheer weight of numbers, while on 17 January 1918 it was reported crowds had raided shops in Tonypandy.

36 See article by Hartshorn in *SWDN*, 29 December 1917.

37 *SWDN*, 14 January 1918.

38 E. May, 'A Question of Control: Social and Industrial Relations in the South Wales Coalfield and the Crisis of Post-War Reconstruction' (PhD dissertation, University of Wales, 1995), pp. 108–41.

39 S. E. Demont, 'Tredegar and Aneurin Bevan: A Society and its Political Articulation, 1890–1929' (PhD dissertation, University of Wales, 1990), Ch. 3.

40 Reulecke, 'Der Erste Weltkrieg', p. 229.

41 Tampke, *Ruhr and Revolution*, p. 54.

42 *Oberbürgermeister* to *Regierungspräsident* Düsseldorf, 7 August 1916, STAD 15984.

43 See *ibid.*, p. 55 and Lützenkirchen, *Der sozialdemokratische Verein für den Reichstagswahlkkreis Dortmund-Hörde*, p. 99.

44 Goch, *Arbeiterbewegung und Arbeiterkultur*, pp. 187–90.

45 D. Egan, 'The Swansea conference of the British Council of Soldiers and Workers Delegates, July 1917: Reactions to the Russian Revolution of February 1917 and the anti-war movement in south Wales', *Llafur*, 1:4 (1975), 12–37.

46 R. Griffiths, *S. O. Davies* (Llandysul: Gomer Press, 1983), pp. 36–8.

47 P. Davies, *A. J. Cook* (Manchester: Manchester University Press, 1987), pp. 21–32; P. Davies, 'The making of A. J. Cook: His development within the south Wales Labour movement, 1900–1924', *Llafur*, 2:3 (1978), 46–8.

48 E. Lucas, *Märzrevolution 1920. Vom Generalstreik gegen den Militärputsch zum bewaffneten Arbeiteraufstand* (Frankfurt: Roter Stern, 1970), p. 23.

49 Witten *Bergamt* to *Oberbergamt*, 22 October 1915, Essen 2 *Bergamt* to *Oberbergamt*, 23 October 1915 and Oberhausen *Bergamt* to *Oberbergamt*, 22 October 1915, STAM 1853.

50 R. Tschirbs, *Tarifpolitik im Ruhrbergbau 1918–1933* (Berlin and New York: Walter de Gruyter, 1986), p. 32.

51 Tenfelde, 'Linksradikale Strömungen in der Ruhrbergarbeiterschaft', p. 223.

52 Tschirbs, *Tarifpolitik im Ruhrbergbau*, pp. 37–46.

53 Arnot, *South Wales Miners*, vol. 2, pp. 152–3.

54 E. May, 'Labour, capital and the state in the south Wales coalfield 1912–1921: the case of Gwaun-cae-Gurwen', in I. Blanchard (ed.), *New Directions in Economic and Social History* (Avonbridge: Newless Press, 1995), pp. 47–55. T.-L. Liu, *The Chameleon State: Global Culture and Policy Shifts in Britain and Germany, 1914–1933* (Oxford: Berghahn, 1999), p. 140.

55 See E. Lucas, *Zwei Formen von Radikalismus in der deutschen Arbeiterbewegung*

(Frankfurt am Main: Roter Stern, 1976), pp. 160–81; Tschirbs, *Tarifpolitik im Ruhrbergbau*, pp. 52–3; Spethmann, *Jahre*, vol. 1, pp. 132–72; Hartewig, *Das unbrechenbare Jahrzehnt*, pp. 247–51.

56 K. F. Gesau, 'Syndikalismus in der Ruhrbergarbeiterschaft zu Beginn der Weimarer Republik, 1918–1925' (Hausarbeit, Münster, 1983), pp. 135–43.

57 H. Rübner, *Freiheit und Brot. Die Freie Arbeiter-Union Deutschlands: Eine Studie zur Geschichte des Anarchosyndikalismus* (Cologne: Libertad Verlag, 1994), pp. 40–1. E. Lucas, *Märzrevolution 1920. Vom Generalstreik gegen den Militärputsch zum bewaffneten Arbeiteraufstand* (Frankfurt: Roter Stern, 1970), p. 58; Gesau, 'Syndikalismus in der Ruhrbergarbeiterschaft', p. 182. The other organisations were the *Allgemeiner Arbeiter Verband, Allgemeine Bergarbeiter Union, Allgemeine Arbeiter Union Essen, Allgemeine Deutsche Bergarbeiter-Union (Düsseldorf).*

58 Gesau, 'Syndikalismus in der Ruhrbergarbeiterschaft', Anhang, p. 121. The *Freie Arbeiter Union Deutschlands* (FAUD), of which the FAU and FVdG were a part, was founded following a conference in Berlin on the 27 and 30 December 1919, Rübner, *Freiheit und Brot*, p. 36.

59 Bock, 'Anarchosyndicalism in the German labour movement', p. 68.

60 *Ibid.*, p. 63.

61 May, 'Labour, capital and the state', p. 55.

62 *Freie Arbeiter Union, Organ der Schacht- und Betriebsorganisation der Hand- und Kopfarbeiter Deutschlands/Wirtschaftsbezirk Rheinland-Westfalen*, 15 September 1921; P. von Oertzen, *Betriebsräte in der Novemberrevolution. Eine politikwissenschaftliche Untersuchung über Ideengehalt und Strukur der betrieblichen und wirtschaftlichen Arbeiterräte in der deutschen Revolution 1918/19* (Berlin/Bonn-Bad Godersberg: J. H. W. Nachf. GmbH, 1976), pp. 212–3; Rübner, *Freiheit und Brot*, p. 42.

63 Gesau, 'Syndikalismus in der Ruhrbergarbeiterschaft', Anhang, p. 123.

64 E. Weitz, *Creating German Communism, 1890–1990. From Popular Protests to Socialist State* (Princeton: Princeton University Press, 1997), p. 151.

65 Gesau, 'Syndikalismus in der Ruhrbergarbeiterschaft', pp. 176–7.

66 *Staatskommissar für die Überwachung der öffentliche Ordnung* to *Wehrkreiskommando* VI, Münster, 16 March 1920, STAM BüK 324.

67 *Polizeiverwaltung* Hamborn to *Regierungspräsident* Düsseldorf, 9 May 1925, STAD 16932.

68 *Staatliche Berginspektion* 3 to *Staatliche Bergwerksdirektion*, Recklinghausen, 7 September 1922, BBA Bestand 32, 4289.

69 *WVZ*, 19 February 1919.

70 J. Tampke, 'The rise and fall of the Essen model, January-February 1919', *IWK*, 13:2 (1997), p. 166.

71 *WVZ*, 10 January 1919.

72 62nd *ordentliche Generalversammlung, Berbgau Verein*, 28 June 1920.

73 *GAZ*, 15 January 1919.

74 *Ibid.*, pp. 160–72; Spethmann, *Jahre*, pp. 218–39.

75 See E. May, 'Charles Stanton and the limits to "patriotic labour"', *WHR*, 18:3 (1996), 145–66, for the attitudes of the Llanhilleth and Fforchaman lodges.

76 Report on Revolutionary Organisations in the UK, Cab 24/87 GT 8036, 21 August 1921.

77 Ministry of Labour, PRO Cab 24/94 CP 285, 10 December 1919.

78 Ministry of Labour, Cab 24/86 GT 8000, 13 August 1919.

79 *Ibid.*, PRO Cab 24/98 CP 686, 19 February 1920 and *ibid.*, PRO Cab 24/94 CP 218, 26 Nov 1919.
80 *LV*, 10 May 1919.
81 Claude Stanfield, Local Labour Leader Study, SWML, AUD 205.
82 Minutes of Proceedings at a meeting between Representatives of Monmouthshire and South Wales Coal Owners' Association and Representatives of the South Wales and Monmouthshire Mechanical and Surface Workers' Union, 27 November 1922, NLW, MV494, pp. 7–8.
83 *LV*, 17 September 1921.
84 There were questions raised about who the union actually represented. See minutes of a meeting between the MSWCOA and the Mechanical Union, 7 December 1921, NLW MV488 and 21 July 1922, NLW MV492. For examples of the campaign and resolutions against the Mechanical Union see *GG*, 23 March 1923; Cambrian Lodge Minutes, 27 March and 8 May 1923, SWCA, MNA/NUM 1/2/20/A4; Oakdale Lodge Minutes, 12 September and 10 May 1921, SWCA, MNA/NUM L/59/1.
85 Burghardt, *Mechanisierung des Ruhrbergbaus*, pp. 253–5.
86 Brüggemeier, *Leben vor Ort*, p. 280.
87 Burghardt, *Mechanisierung des Ruhrbergbaus*, p. 289.
88 *Bergrat* Funcke to Dr Jüngst, 5 October 1925, BBA 625.
89 Graf Alexander Stenbock-Fermor, *Meine Erlebnisse als Bergarbeiter* (Stuttgart: J Engelhorns, 1928), pp. 44–5.
90 Ogmore and Garw UDC Minutes, 18 October and 8 November 1918, GRO UG/OG 14.
91 Coombes, *Poor Hands*, p. 103.
92 T. Boyns, 'Jigging and shaking: technical choice in the South Wales coal industry between the wars', *WHR*, 17:2 (1994), 230–50.
93 See Ch. 2 for Tenfelde's argument.
94 Weitz, *Creating German Communism*, p. 121.
95 On rough and respectable workers in Britain see E. Hobsbawm, *Worlds of Labour: Further Studies in the History of Labour* (London: Weidenfeld and Nicolson, 1984), Ch. 13. On the lessening stigma attached to relief see J. E. Cronin, *Labour and Society in Britain 1918–1979* (London: Batsford Academic and Educational, 1984), p. 82.
96 Llanbradach Workmen's Hall, Minute Book, 7 October 1921, GRO D/D X 816.
97 *Ibid.*, 13 June 1923, GRO D/D X 816.
98 R. J. Shearer, 'Shelter from the storm: politics, production and the housing crisis in the Ruhr coal fields, 1918–24', *JCH*, 34:1 (1999), 30–1.
99 Tenfelde, 'Linksradikale Strömungen in der Ruhrbergarbeiterschaft', pp. 220–3. See also K. Tenfelde, 'Die Bergarbeiter, ihre Gewerkschaften und der Kapp-Putsch', in J. Gorlas and D. J. K. Peukert, *Ruhrkampf 1920* (Essen: Klartext, 1986), pp. 45–8.
100 Lucas, *Zwei*, pp. 280–1. Remscheid is a city outside the Ruhr, whose history stretches back to the Middle Ages.
101 Dortmund 1 *Bergamt* to *Oberbergamt*, 14 January 1919 and Recklinghausen East *Bergamt* to *Oberbergamt*, 11 January 1919, STAM OBA 1160.
102 Dortmund 3 *Bergamt* to *Oberbergamt*, 3 June 1919, STAM OBA 1795 and Hamm *Bergamt* to OBA, 3 June 1919, STAM OBA 1795.
103 *Polizeidirektor* Bochum to *Landrat* Bochum, 7 May 1922, STAB LA 1315.
104 Kleßmann, *Bergarbeiter*, p. 191.
105 *Ibid.*, pp. 116–21.
106 See reports on various meetings of the *Polnische Arbeiterpartei* in STAM Reg. Arnsberg 14054.

107 Jäger, *Bergarbeitermilieus*, p. 154.
108 F. J. Brüggemeier, '"Volle Kost volle". Die Wohnungsverhältnisse der Bergleute an der Ruhr um die Jahrhundertwende', in Hans Mommsen and Ulrich Borsdorf, *Glück auf, Kameraden. Die Bergarbeiter und ihre Organisationen in Deutschland* (Cologne: Bund-Verlag, 1979), pp. 172–3.
109 D. Hopkin, 'Social reactions to economic change', in T. Herbert and G. E. Jones, *Wales Between the Wars* (Cardiff: University of Wales Press, 1988), pp. 53–5.
110 Pope, *Building Jerusalem*, pp. 224–5.
111 Henry Lewis, Abercraf Community Study, SWML AUD 201.
112 Pope, *Building Jerusalem*, pp. 241–9.
113 B. Hogenkamp, 'Miners' cinemas in South Wales in the 1920s and 1930s', *Llafur*, 4:2 (1985), 65; Stead, 'Wales and film', in Herbert and Jones (eds), *Wales Between the Wars*, p. 161.
114 M. Johnes, 'That Other Game', p. 228.
115 Ammanford Rugby Club, Sixty Years of Rugby 1881–1947: Ammanford R. F. C. Diamond Jubilee, 1947, p. 29.
116 H. Francis, 'The Anthracite Strike and the disturbances of 1925', *Llafur*, 1:2 (1973), 53–66; Church and Outram, *Strikes and Solidarity*, pp. 122–7.
117 S.Gehrmann, 'Fußball in einer Industrieregion. Das Beispiel F. C. Schalke 04', in Reulecke and Weber (eds), *Fabrik, Familie, Feierabend*, pp. 385–98.
118 Report on Revolutionary Organisations in the UK, 25 September 1919, PRO Cab 24/89 GT8228.
119 V. Schmidtchen, 'Arbeitersport – Erziehung zum sozialistischen Menschen? Leitwiese und Jugendarbeit in zwei Ruhrgebietsvereinen in der Weimarer Republik', in Reulecke and Weber (eds), *Fabrik, Familie, Feierabend*, pp. 345–76.
120 *Regierungspräsident* Düsseldorf to Minister of the Interior, 11 February 1925, STAD 16910.
121 Abrams, *Workers' Culture in Imperial Germany*, p. 171.
122 Coombes, *Poor Hands*, pp. 144–7; *AVC*, 2 June 1921.
123 Rees Davies, Abercraf Community Study, SMWL, AUD 207.
124 *GG*, 19 November 1926.
125 For the various forms of protest and survival in the Ruhr see Hartewig, *Das unbrechenbare Jahrzehnt*, pp. 215–76. Looting of shops was more a feature of the turbulent Ruhr. In Essen, for example, miners raided the city slaughter house, *Oberbürgermeister* Essen to *Regierungspräsident* Arnsberg, 1 October 1923, STAD 16845. See other examples in H. Spethmanns' three-volume work, *Zwölf Jahre Ruhrbergbau. Aus seiner Geschichte von Kriegsanfang bis zum Franzosenabmarchen 1914–1925* (Berlin: Reimar Hobbing, 1929/30) and Stenbock-Fermor, *Meine Erlebnisse als Bergarbeiter*. The working of outcrop coal seems to have been more prevalent in south Wales, probably due to its particular geology. However, it was not always looked upon favourably by strike leaders. See Aneurin Bevan's robust reaction in M. Foot, *Aneurin Bevan: A Biography 1897–1945* (London: MacGibbon and Kee, 1962), p. 73. Organised entertainment and arrangement of alternate work during the strikes was widely reported in the press. See, for example, *AVC*, 9 June and 16 June1921 respectively.
126 Francis and Smith, *The Fed*, pp. 54–69.
127 Gilbert, *Class, Community, and Collective Action*.
128 Foot, *Aneurin Bevan*, pp. 50–62.
129 Ferndale Lodge Minutes, 4 June 1920, SWCC MNA/NUM L/33/A2.

130 'Relatively' because the actions of some labour leaders, such as the miners' leader Frank Hodges and railwaymens' leader J. H. Thomas, were vilified for their roles in the 1921 and 1926 strikes.
131 Weitz, *Creating German Communism*, pp. 152–3.
132 R. McKibbin, *The Ideologies of Class: Social Relations in Britain 1880–1950* (Oxford: Clarendon Press, 1990), Ch. 3; Duncan Tanner, 'The parliamentary electoral system, the "Fourth" Reform Act and the rise of Labour in England and Wales', *Bulletin of the Institute of Historical Research*, 1983, 205–19.
133 *CWM*, January 1923, pp. 9–10.
134 See *ibid.*, March 1923 and September 1923.
135 *Ibid.*, January 1923, p. 10.
136 *CJ*, 20 February 1920; Ammanford, UDC Minutes, 7 April 1919, CCRO. *AVC*, 7 April 1921; Ammanford UDC Minutes, 5 April 1921, CCRO.
137 *AVC*, 3 September 1925.
138 Ferndale Lodge Minutes, 21 February 1922, SWCC MNA/NUM L/33/A4.
139 May, 'Charles Stanton', pp. 145–66.
140 See M. Lieven, 'A fractured working-class consciousness? The case of the Lady Windsor Colliery Lodge', *WHR*, 4:21 (2003), 729–56.
141 DDR historian Otto Hennicke describes the political breakdown of the Ruhr Red Army as KPD 30.9%, USPD 58.4% and SPD 10.7%. Hennicke, *Die Rote Ruhrarmee*, (Berlin: Ministry for National Defence, 1956), p. 54.
142 H. Spethmann, *Die Rote Armee an Ruhr und Rhein. Aus Kapptagen* (Berlin: Reimar Hobbing, 1930), pp. 50–9.
143 Tschirbs, *Tarifpolitik im Ruhrbergbau*, p. 101.
144 Tenfelde, 'Bergarbeiter', in Gorlas and Peukert (eds), *Ruhrkampf 1920*, pp. 42–55.
145 E. Lucas, *Märzrevolution 1920. Der bewaffnete Arbeiteraufstand im Ruhrgebiet in seiner inneren Struktur und in seinem Verhältnis zu den Klassenkämpfen in den vershiedenen Regionen des Reiches*, vol. 2 (Frankfurt: Roter Stern, 1973), pp. 72–3.
146 Third Kav. Division, 29 April 1920 and 22 April 1920, STAM BüK 179.
147 *Ibid.*, pp. 78–9.
148 *Oberpräsident* to Minister of the Interior, 10 October 1924, STAD 16910. For a detailed analysis of the French Occupation of the Ruhr and ensuing crisis see C. Fischer, *The Ruhr Crisis, 1923–1924* (Oxford: Oxford University Press, 2003).
149 KPD Rundschreiben, 3 June and 19 September 1924, STAD 16934. Weitz, *Creating German Communism*, Ch. 5.
150 *Polizeipräsident* Essen to *Regierungspräsident* Arnsberg, 7 April 1926, STAD 16934.
151 D. Rees, 'Morgan Jones, educationalist and Labour politician', *Morgannwg*, 31 (1987), 66–83.
152 A. Horner, *Incorrigible Rebel* (London: MacGibbon and Kee, 1960), p. 50.
153 *WM*, 15 November 1920.
154 *LV*, 28 December 1918.
155 *AVC*, 17 September 1925.

# 5

# Miners' trade unionism and political identities, 1918–1926

He used to think that Liberalism stood for that once [the welfare of the majority of the people]. When he first took an interest in politics, the great voices of Gladstone, Tom Ellis and the young Lloyd George still rang ... In those days he felt like Wordsworth that it was a blessing to be alive, but; 'Oh! For a touch of the vanished hand/And the sound of that voice which is/ still.' They did not find the idealistic note there was in those days. Regretfully he said it [the Liberal Party] was a hard, commercialised party. He had ceased to hope for the coming of the golden age he longed and yearned and prayed for.

Revd Waldo Lewis, explaining his conversion to Labour during electioneering at Carmarthen.[1]

We're cheated by everyone, but by the KPD the least, that's why we vote for them.

A miner's comment to Stenbock-Fermor during his sojourn in the mines.[2]

## Introduction

The preceding chapter sought to illustrate the effects that total war had upon the lifeworld and civil society in the south Wales and Ruhr coalfields. This chapter turns back to the discursive practices of the trade unions and the political parties as they sought to interpret the turbulent events in which they found themselves enmeshed. In south Wales, the Fed was strengthened, while the Labour Party eventually superseded the Liberals. In the Ruhr, new syndicalist-communist organisations challenged the older unions. The SPD vote, following the high-point of the 1919 *Nationalversammlung* campaign, was devastated in the subsequent Reichstag elections. However, the question remains as to how this intensification of homogenisation and fragmentation consolidated through the discourse of the organisations, both industrial and political. How was the Labour Party able to usurp the position of the Liberals in the mining valleys, while the Fed amalgamated the smaller craft unions immediately after the war? On the other hand, why did the socialist labour movement in the Ruhr split between the Communists and the Social Democrats?

In post-war south Wales, the Labour Party continued to incorporate elements of Liberalism, such as concerns with disestablishment and nationalism, into its discourse, while at the same time strengthening its image as the party of the working class during the strikes and lockouts. In the context of conflict between labour and capital in the 1920s, the party was thus able to adopt the mantle of progressivism from Liberalism. In the Ruhr, the Centre Party largely continued to propound a discourse similar to that used in Wilhelmine Germany, although it dropped its monarchism. The SPD meanwhile found itself in the difficult position of being the party of the Republic. The USPD, and later the KPD, partly inherited from the SPD the mantle of the anti-capitalist party. The SPD, therefore, found its position as the party of the left seriously challenged. This was a situation that had not existed under the Empire. If, as Ross McKibbin has suggested, 1920s British politics was essentially the extension of an ossified Edwardian form of politics,[3] then Weimar politics represented new and turbulent territory.

## Socialisation and nationalisation

As pointed out in the previous chapter the issue of socialisation was symptomatic of the divisions in miners' trade unionism in the Ruhr. The term became a slogan exploited by all the trade unions, yet beneath the slogans very different conceptions and values were attached to the idea of socialisation. Although several variants and numerous plans existed, three broad approaches can be distinguished: socialisation as nationalisation, socialisation through small share ownership and socialisation as workers' control.

The fragmentary nature of the various radical trade unions in the Ruhr makes identifying a common approach to workers' control of the mines difficult. The cases of 'wild' socialisation in 1919 have been portrayed as a direct response to economic difficulties, with little to do with a loftier sounding, but less materialistic, growth of class consciousness.[4] This view is certainly justified when applied to the 'wild' socialisation. However, the Commission of Nine, the body elected following the seizure of the *Zechenverband* offices in 1919, did attempt to provide an ideological basis for socialisation. Its particular conception of socialisation was set out in the pamphlet *Die Sozialisierung des Bergbaues und der Generalstreik im rheinisch-westfälischen Industriegebiet*. The Commission was careful to draw a distinction between its idea of socialisation and the 'wild' socialisation. The establishment of Workers' and Soldiers' Councils was presented as a positive step towards socialisation, but 'unfortunately confusion still reigned on how socialisation was actually to be understood'. The pamphlet argued that socialisation did not simply mean 'the pit belongs to us' and that coal could be exchanged for food supplies. This was 'childishness', propagated by the syndicalists.[5] Instead the miners would 'soon come [to] realise that control of the mining works by the workers must precede socialisation, that it is a prerequisite of socialisation and at the same time its important component. And not only control of the individual works, but also the employers' organisations.'[6]

The Commission, therefore, perceived socialisation as meaning workers' control of not only the mines, but also the whole administrative apparatus of the industry. The Commission modelled its socialisation plans on the *Rätesystem* (council system) established during the Revolution and in the minds of the Commission the two were inseparably bound together.[7] The Commission explained the *Rätesystem* as:

> the workers, the industrial workers and white collar workers (*Angestellten*), the agricultural workers and the farmers, the soldiers, administrating their affairs themselves. It means, in its full application the complete transformation of the state, in which the rule of bureaucracy and parliamentarianism is to a great degree superfluous. Its partial application means an advance against capital's power and bureaucracy, in which control of the capitalist works and authorities is brought about – not through Parliament, whose control becomes completely illusionary in the course of time because of the complication (*Komplizierung*) of the state machine – but by the workers.[8]

The system was made up of a pyramid of councils divided into four levels:

1 *Steigerrevierräte*, which was responsible for safety in the individual districts in the mine. Echoing the *Alter Verband*'s pre-war concern with the safety-men, and thereby illustrating some of the continuities between the organisations, the Commission argued that these new councils would not prove a 'white ointment', but would be able to properly enforce health and safety regulations;
2 *Zechenräte*, which oversaw the whole mine;
3 *Bergrevierräte*, which represented the common interests of mines in certain geographical areas;
4 *Zentralzechenrat* (Central Mines Council), which was to oversee the entire condition of the industry in the region.[9]

Interestingly, it was suggested that the Central Council would have the toughest job as it would have to watch over the various employers' organisations, suggesting that those organisations would still exist. However, the scheme was seen very much as part of a process working towards workers' control and ultimately there would be no role left for the owners.

The slow introduction of socialisation further distinguished the Commission from the syndicalists. The Commission sought to portray the syndicalists as simple or unreasonable. All 'thinking workers' were to resist efforts at wild socialisation, while 'no reasonable person believed that socialisation could be immediately enforced'. The councils were to exercise control as a check to the encroachment of the mineowners' power. 'The miners are ready to tolerate the yoke of private capitalism a while [longer], but they insist that the right of control, which is in their interest and in the interest of the commonality absolutely necessary, must be immediately introduced. The councils shall serve this control and only this control.'[10]

If the council system represented a transitional stage, a milestone on the way to socialisation, the Commission was in no doubt of the legitimacy of its demands. In response to the Government's refusal to recognise the Commission, the pamphlet stated:

> The gentlemen of the Government can indeed accept or reject the authority of this Commission. That is their affair. But they cannot decide, whether it continues to exist or not. **The Commission of Nine is an organ of the revolutionary workforce.** They have created this Commission and alone decide [its future]; the Government has nothing at all to say and the workforce can cry: Hands off! The revolutionary workforce itself has created this organ to prepare the socialisation of the mining industry. This socialisation will come; whether it is enforced by the Government or against it that is still the question. In any case it will be enforced by the workforce ... **So long as the miners of the Rhenish-Westphalian industrial area believe that [the industry] must be socialised, they will require an organ that prepares this socialisation and only people who have the full trust of the revolutionary workers can form it.**[11]

For the Commission its legitimacy was located in the perceived revolutionary nature of the working class, and of the miners in particular. The Commission itself stood outside the normal structures of authority. Indeed, the insistence that the elected Government had no authority over the Commission raised the issue of dual control. The Commission clearly perceived itself as the nucleus of a new society: a society that would be realised through the transformation of the existing order. It has been suggested that the *Rätesystem*, of which socialisation was an integral part, offered a third way between communist dictatorship and bourgeois democracy.[12] However, the Commission too placed the interests of the working class above the more diverse social interests represented by the *Nationalversammlung*. Such heavy emphasis on the revolutionary working class, coupled with the radical left-wing political composition of the Commission after 16 February 1919, was unlikely to expand the appeal of socialisation beyond the working class, or beyond even a certain section of it. This was especially so given the existence of competing conceptions of socialisation. It therefore seems unlikely that the issue of socialisation could have attracted broad enough support among the working class, let alone across German society as a whole, to provide a viable 'third way'.

Yet if the Commission rejected the authority of the SPD-dominated government, it was conscious that its theoretical basis was still to be found within social-democratic ideology. Criticism of SPD ministers was harsh and the Commission felt compelled to point out that, in its original form, it had included SPD representatives.[13] In fact, the Commission played upon its social-democratic heritage.

> Until now socialisation was for every Social Democrat a clear and unambiguous concept. It meant 'expropriation of the expropriators', as Karl Marx classically expressed it. It means: revocation of private property in the means of production, expropriation of the owner, the transfer of this means of production to the ownership of society, to the communality. The ruling Social Democrats appear with the acceptance of their offices to have thrown this social-democratic view over board,

> because in their ordinance socialisation is synonymous merely with 'influence through the *Reich*' and a 'share of the people's community (*Volksgesamtheit*) in its profits'. Instead of the social-democratic, robust view, 'socio-political' blather! Instead of the struggle against the power of capital, a shameful pact with the capitalist rulers![14]

In this sense the Commission members, and by implication their wider supporters, were being portrayed as the keepers of the old faith as adopted by the Erfurt conference of 1891. The MSPD, especially those that had accepted office, were seen at best as individuals who had lost their way, at worst traitors. Later the pamphlet described the *Alter Verband*, *Gewerkverein* and ZZP decision not to support the call for a general strike in February 1919 as a betrayal.[15] The *Alter Verband* was singled out for special criticism. 'Once – long, long ago – [the *Alter Verband*] carried forward the red pennant of the fighting proletariat, today its leaders have made it into a strike-breaking organisation.' Such language echoed pre-war accusations directed at the *Gewerkverein* by the *Alter Verband*. Expanding the betrayal/faith metaphor the *Alter Verband* leaders were described as 'genuine Judases'.[16] For the Commission the MSPD and *Alter Verband* were no longer the vanguard of the class struggle. That honour now fell to the Commission.

The conclusion of the pamphlet clearly distinguished the Commission's idea of socialisation from others'. 'Socialisation [was] not mere nationalisation (*Verstaatlichung*) of the mining works, but the transfer of mining property to the community and the administration of the industry to the workers' councils.'[17] Ultimately, then, control was to rest with the workers themselves, the state playing only a subsidiary role.

Although the primary evidence is somewhat fragmentary, it is clear that the diverse syndicalist-communist trade unions also supported a socialisation predicated relentless class conflict. In fact, the antagonistic attitude towards employers was even more clearly expressed by these organisations. The *Union der Hand- und Kopfarbeiter*, despite tensions between syndicalist and communist wings, was clear on the relationship between employers and employed. The *Union* claimed in its organ that 'resistance to socialisation is grounded in the class instinct of the exploiters which is nurtured and maintained by the interests of property'. On the other hand, 'socialisation itself is sustained by the class interest of the exploited [and] is dictated by inalienable, natural human rights. Measured against the strengths of these two camps, [the realisation of] socialisation means an extremely hard and stubborn struggle'.[18]

The emphasis on class, struggle and betrayal also provided the main themes within the discourse of the other syndicalist-communist unions and left-radical parties. An *Allgemeine Arbeiter Union* (AAU) leaflet distributed at the *General* mine called on the 'working class [to] take up the struggle against the whole of capitalist society, overthrow the ruling authority and erect [their] own power'. The workers of the Ruhr, miners and metalworkers were now the 'vanguard' of the West European revolutionary workers' movement. 'They must see the clearest, travel the furthest down the route to revolution; they must most keenly recognise their enemies.' Among those enemies were the other trade unions. The FAUD were criticised for not accepting the importance of the dictatorship of the proletariat, while the FAUG was seen as not accepting the basic nature of the struggle. True revolutionaries were to leave these bodies and join the AAU, for the only way the workers could realise their true freedom was through 'uncompromising class conflict.' For the AAU, like the *Union*, there was no middle ground in this conflict. It was a question of being for or against the revolution.[19]

The syndicalist FAUD also believed that the other unions needed to be fought as they were 'after all only institutions of bureaucracy. Where a bureaucracy rule little is achieved, therefore our task must be to educate and enlighten the miners.' On the other hand, an invitation to affiliate with the Communists was rejected since the FAUD was 'a purely economic fighting organisation and not a political organ'.[20] Parliamentarianism and political parties were also condemned because they were tainted by capital.[21] A pamphlet entitled 'What do the Syndicalists want?' differentiated the movement from the 'Marxist parties'. The SPD, USPD and KPD were supposedly obsessed with the conquest of political power, which drew attention away from what really mattered, the economic struggle.[22] What was lacking in German society was a spirit of solidarity among the working class. It could be created through a general strike.[23]

It is clear then that there were differences in the discourses of the communist and syndicalist groups. At the same time the boundaries between these discourses were permeable, especially immediately after the Revolution. Both sides published leaflets and booklets in an effort to distinguish their organisational identity.[24] Yet the various left-wing trade unions, be they syndicalist or communist, shared a common view of industrial relations as invariably combative. Any system of bargaining was therefore seen as an unwelcome compromise, a 'stale peace between capital and labour'.[25] Where the unions differed was in how the conflict was to be resolved. All agreed on the importance of the strike as a weapon but failed to agree on the role of political action. The FAUD saw the true site of struggle as the pit itself. The miner was seen as fighting on one front. He was an economic warrior pure and simple. Too much engagement with politics could only provide a distraction and prove a weakness. For the Communist trade union members, the strike was an adjunct to political action. Ultimately, it was the establishment of a dictatorship of the proletariat that would free the workers. The miner was the 'vanguard' of the revolutionary workforce in both the industrial and political spheres. The general strike, the armed insurrection and the ballot were all weapons in the armoury of the proletariat, their usefulness fluctuating according to circumstance.

Despite these differences, the similarities between the language used by the left-wing unions and that used by the *Alter Verband* before the war are striking. The relationship between employers and employed is defined in terms of conflict and socialisation reduced to a trial of strength. Indeed, the emphasis and importance placed on the strike as a weapon echoes the belligerent attitude of the *Alter Verband* during the early 1890s, before Möller and Hue forged a more restrained image.[26] In some respects the language employed by the syndicalist and communist unionists represented a reassertion of that more aggressive strain in the pre-war discourse of the *Alter Verband*, complemented by the re-emergence of the tradition of pit-head meetings and direct democracy. As described above, the USPD, KPD and syndicalists often drew their strength from the old left-wing currents of social democracy and were conscious of their social-democratic heritage.

In much the same way the Communists regarded themselves as the true inheritors of the revolutionary tradition. The old link with social democracy was not forgotten and could be shown in a positive light. For example, during the 1925 presidential election a potted biography of the Communist candidate and KPD leader, Ernst Thälmann appeared in the *Westfälischer Kämpfer*. Like pre-war SPD propaganda it concentrated on the fact that Thälmann was a working-class candidate. Among his credentials was his membership of the SPD from the age of sixteen. His break with the SPD occurred during the war when he challenged the 'treacherous policy of the Social Democrats'. Yet his previous party membership was still included as part of his '20 years in the labour movement'.[27] The idea of treachery implied that, unlike the remaining SPD activists, Thälmann had kept the faith. During the campaign Thälmann spoke at both Essen and Dortmund. Symbolising the internationalism of the movement unnamed representatives of both the English and French Communists spoke at his meetings. Thälmann himself stressed how the KPD had taken the mantle of defenders of the working class. He conflated the Republic and the monarchy and depicted them as united against the working class. Worse, those who ruled in the Republic were also those who had run business under the monarchy. The blame was laid squarely on the shoulders of the SPD. Even here in the Ruhr, 'the historic land of the German class struggle', the SPD had betrayed the workers in the aftermath of the Kapp *Putsch*.[28]

The idea that the SPD had betrayed the revolutionary workers and were a spent force was a constant theme within KPD discourse during this period. This was often represented as an alliance between the SPD and the bourgeoisie.[29] No distinction was made between the Republic and the old regime. One pamphlet posed the question, 'Ebert or Ludendorff?', but suggested that there was no difference. 'One year of the middle-class Republic has shown you that it is nothing more than the old dictatorship of capital.'[30] Unsurprisingly, leaflet literature advocated revolution instead of Republic. Seizing on the enthusiasm of revolution, early leaflets contrasted the Council Republic (*Räterepublik*) with the middle-class Republic proposed by the SPD. This was often bound up with the issues of workers' control and socialisation.[31] Many pamphlets also had an internationalist tone, calling on the workers of the world to unite in defence of Soviet Russia.[32]

In contrast to left-wing parties and unions, the *Alter Verband*, whose ideological approach was still very much shaped by Hue until his death in 1922, regarded socialisation as something akin to nationalisation in Britain. It is important to note here that the *Alter Verband* characterised its conception of socialisation as *Vergesellschaftung* (socialisation), rather than *Verstaatlichung* (nationalisation). In fact Hue, like the Commission of Nine, rejected the latter, although he admitted that there were many Social Democrats who had not realised that the union's demand was not *Verstaatlichung.*[33] However, the *Bergarbeiterzeitung* commented in 1919 that previously the union had used the word nationalisation, rather than socialisation, as it was better understood. In fact, nationalisation was a goal it had sought since its foundation.[34]

Nationalisation was therefore an issue discussed by both the MFGB and the *Alter Verband* before 1919. The MFGB had proposed it in 1912, while during the war an article had appeared in the *Bergarbeiterzeitung* promoting the idea of nationalisation and refuting the argument that state mines were less profitable than their private counterparts.[35] In fact, a common tactic in the debate over nationalisation, both in Germany and Britain, was to compare the position of the existing German state mines with the private mines in the Ruhr.[36]

The MFGB 1919 proposals for nationalisation, however, differed from the 1912 plan in that control of the industry would lie with both the workers and the state.[37] Initiated by the SWMF, the proposals called for the 'transfer of the entire industry from Private Ownership and Control to State Ownership with joint control and administration by the Workmen and the State'. The miners were not to be the passive subjects of nationalisation, but active in the management of the mines. This point was further elaborated by the former SWMF Executive member and current MFGB Secretary, Frank Hodges, who stated:

> The principal point to me in this resolution is one that has not yet been elaborated, that is the question of control. For the last two or three years a new movement has sprung up in the labour world which deals with the question of joint control of the industry by representatives from the side which represents, for the most part, the consumer and representatives of the workmen who are producers. Nationalisation in the old sense is no longer attractive. As a matter of fact you can have nationalisation but still be in no better position than you are now under private ownership. That is the experience of institutions which have been State owned and State controlled for many years ... Now is it any good to have these mines nationalised unless we are going to exercise some form of control as producers? If not, the whole tendency will be towards the power of the bureaucracy. We shall be given no status at all in the industry, except to be the mere producers, as we have been in the past years. Under State ownership the workmen should be desirous of having something more than the mere question of wages or mere consideration of employment; the workmen should have some directive power in the industry in which they are engaged.[38]

The SWMF proposals envisaged a continuation and extension of the *de facto* bipartite system of industrial relations established during the war.[39] Ultimately there was no place for private capital within this system. The new rules adopted by the SWMF in 1917 declared its goal to be the 'complete abolition of capitalism'.[40]

The argument for nationalisation was put most succinctly in Hodges's book, *Nationalisation of the Mines.* The book is symptomatic of the confidence which imbued SWMF publications and meetings immediately after the war. Nationalisation, it was repeatedly claimed, was almost inevitable. Hartshorn, for example, commented that 'there is no question of whether the Government will agree to nationalisation or not; the adoption of that principle in the very near future is absolutely inevitable'.[41] Not only was nationalisation inevitable, but it was also right, for the working class had now reached such a stage of education that it could take over the direction of the industry. This educated working class yearned 'for the status of responsible manhood in industry and rebels against any cramping institution which thwarts its aspirations for freedom'.[42] The image of the educated, responsible miner permeated Hodges's book. Improved education, he claimed, had made the notion that labour was a simple commodity untenable. Instead, Hodges claimed:

> The labourer, besides having labour power to sell is also possessed of a human soul which is feeling the urge of strong aspirations ... He wants a greater share in the direction of industry. The miner wants to be in his job as a complete human being ... His new status will be that of a partner in industry. Manual labourer and technician will engage together in the mine, and sell their commodity, labour, not to a non-labouring shareholder, but to all men in their calling, as their contribution to the production of the finished article.[43]

This new status would mean new responsibilities, which would encourage an even greater sense of social responsibility among the miners.[44] This emphasis on responsibility, education and partnership would superficially seem to echo the discourse of a pre-1910 generation of miners' leaders. However, the identity that Hodges was trying to create differed from Mabon's in that the miners were in partnership with other workers rather than employers. While the labourer and technician are present in the above extract, the owner is conspicuous by his absence. Earlier in the book Hodges denied that there could ever be an identity of interest between the employers and the workers. Pointing to the increasing number of working days lost to disputes between 1916 and 1919, Hodges argued that there could be 'no co-operation between workmen and employers; there is no common motive in industry; no conscious attempt to attain the highest efficiency; no elimination of waste; no internal economies. Both pursue opposite aims.' Moreover, in direct contrast to Mabon, Hodges denied the owners' right to profit.[45] The very existence of profit could only ensure continuous industrial unrest. Much like the discourse of the *Alter Verband* and post-war left-wing unions, Hodges accepted that relations between owners and workers could only ever be understood in terms of struggle.

In a subsequent chapter Hodges made it clear that the aspirations of the miners could only be met when the obstacle of capitalist influence was removed.[46] But the miner was not only to find individual and class realisation through nationalisation. He was also to act as a model for other workers. Just as *The Miners' Next Step* had sought to portray the miners as the vanguard of the British working class, so too Hodges envisaged them as setting an example for other workers to follow once nationalisation was achieved.

It is clear from Hodges's book that the more collaborationist trend within the

SWMF discourse, weakened during the war, was submerged by a more aggressive tone in the post-war period. By the post-war period, Hodges was not regarded as being on the extreme left of the movement, although in the pre-war years he had been an ILP member and was briefly identified with Ablett and the URC.[47] At the February 1919 MFGB conference in London, for example, more radical members of the South Wales delegation such as Noah Ablett and S. O. Davies attacked the findings of the Sankey Commission, while Hartshorn and Hodges supported it.[48] Moreover, looking back over his career following the vilification of his role in the 1921 strike, and five years after the publication of *Nationalisation of the Mines*, he claimed that he always sought to avoid conflict and preferred negotiation to strike action.[49] The book is illustrative of how acceptance of nationalisation and removal of capitalist influence became the mainstream position in the SWMF and provides a good example of how the more radical, combative identity of the miners, succoured by wartime industrial relations, began to marginalise more co-operative conceptions.

Meanwhile, in the Ruhr, the unrest and attempts at 'wild' socialisation initially brought the leaders of the various unions and political parties together. At first the *Alter Verband* organ seemed supportive of the proposed council system, although critical of the USPD and Spartakists' lack of patience.[50] However, following the break between the left-wing parties in 1919 the *Alter Verband* propagated its own plan for socialisation in opposition to the radical unions. Hue, like Hodges, was keen to justify socialisation on the grounds that it would make the industry more efficient. The majority scheme produced by the Socialisation Commission in 1919, to which Hue was a signatory, proposed that the industry be so reorganised that it was put in the service of the community rather than profit, thereby ending the exploitation of the miners and the consumers. The ownership of existent state and private mines would be transferred to a *Deutsche Kohlengemeinschaft* (German Coal Community). This was headed by a five-man directorate elected from the *Reichskohlenrat* (Reich Coal Council) which was made up of representative of the Government, owners, workers and consumers.[51]

Clearly, Hue's conception of socialisation differed somewhat from Hodges's plan for nationalisation. The concern to involve all sections of the industry as well as government and the consumers must be seen in terms of the socio-political background in which the *Alter Verband* was operating. First, the *Alter Verband* was in a very much weaker position than the SWMF. Not only was its influence on industrial relations very weak, but its organisation was threatened by the new radical unions. The fragmented nature of unionism meant that even if joint government-worker administration was introduced, the *Alter Verband* was not guaranteed to be the sole or even main representative of the miners' side. A more far-reaching plan might also hand the initiative to the syndicalists or the KPD and make it difficult to differentiate the *Alter Verband*, which was already losing members, from the other unions. The SPD also found much of its electoral support eroded following the 1920 Reichstag election. Secondly, the *Alter Verband* leaders as Social Democrats recognised the precarious position of the SPD-led Government. Not only were they aware of the needs of reconstruction, they were also undoubtedly wary that any attempt to implement a more radical scheme might upset the government's precarious relations with German military and the Allies.

Despite these differences, there are some striking similarities in the discourse used by SWMF/Labour and *the Alter Verband*/SPD. Nationalisation in both regions was presented as not solely the miners' problem, but one for the nation. The MFGB publicity campaign, for example, was entitled 'Mines for the Nation' and appealed directly to other groups in British society. The *Alter Verband* and SPD discourse, reflecting the problems facing both the party and union, emphasised the national dimension even more than the SWMF. *Alter Verband* activist Hermann Löffler placed socialisation within the context of the miners' right to co-determination in the industry. Socialisation was portrayed as the culmination of a path that had begun with the foundation of the union and passed through the introduction of workers' committees and safetymen. Löffler emphasised that socialisation would place the industry in the service of the nation and democracy.[52] Pointing to Russia, Löffler argued that the Communist variant of socialisation had destroyed the economy and brought foreign military forces on to Russian soil. This type of socialisation was rejected by the Social Democrats in the interests of the German people (*Volk*). The socialisation of the coal industry was not meant to 'serve an occupation or a class, but the people's community (*Volksgemeinschaft*)'.[53]

Moreover, although the MFGB slogan 'Mines for the Nation' may have not been replicated by the *Alter Verband*, Hue did stress in his *Die Sozialisierung der Kohlenwirtschaft* that the socialisation was not simply a workers' question, but one for the wider community. Yet underpinning this appeal was a similar idea that the miners would be able to achieve their full potential through socialisation. Hue seized on Imbusch's words to expand on the idea that humanisation of work was the end goal of socialisation. 'The economy should in reality serve humanity. We do not live merely to produce; we produce in order to live.'[54] This conscious use of *Gewerkverein* discourse was partly due to closer relations between the two unions through their co-operation in the ZAG. However, Hue also aimed at bringing the *Gewerkverein* on side and cited Imbusch several times in the pamphlet in order to legitimise his argument. Hue's references to the Christian trade unionist were a continuation of a trend evident in *Alter Verband* discourse before the war of emphasising the similarities in bread and butter demands in order to strengthen the case for amalgamation.

Although relations between the *Alter Verband* and the *Gewerkverein* had improved during the war, the *Gewerkverein* consistently protected its own organisational identity and any suggestions of amalgamation were still met with a blunt negative. Echoing Imbusch's 1906 pamphlet, an article entitled 'Is an amalgamation of the miners' unions possible?' appeared in *Der Bergknappe* in 1918. The demand for amalgamation allegedly originated from 'irresponsible radical agitators' and 'idealists'. Just like Imbusch's earlier work, the article looked back over the history of the two unions. Again it was their different world-views that distinguished the unions and made amalgamation impossible and 'in the last decade the fundamental positions of the different trade union movements [had] not changed'. A strong Christian union was required to counterbalance the materialism of the Free Trade Unions. Drawing on fears of Bolshevism, the article suggested that the lack of such a balance could tip the SPD into communism and the *Alter Verband* itself could come under the influence of radicals. On the other hand, cooperation on practical occupational questions was desirable. In fact, the question of amalgamation was seen as undermining cooperation as it destroyed trust between the unions.[55] Subsequent articles took a similar line. The *Gewerkverein* was continually represented as a Christian movement, its religious nature contrasted to the materialistic and socialistic qualities of the *Alter Verband*.[56] This sense of continuity in the values of the *Gewerkverein* was further reinforced at their annual conferences. Speaking on the goals of the union, Imbusch stated that their attitude to the employers remained 'in the new Germany as in the old'.[57]

Despite the commitment to cooperation, the relationship between the *Gewerkverein* and the *Alter Verband* was often strained, and occasionally conflict broke out. For example, the language used during the 1922 strike was reminiscent of pre-war inter-union relations. While the *Alter Verband* supported the strike, the *Gewerkverein* rejected it. The *Bergknappe* carried the headline 'we are opponents of the revolutionary class struggle' and reprinted a speech given by the *Gewerkverein* youth leader, Wiedfeld, at a conference in Bochum.

> The social-democratic trade union movement stands on the basis of the revolutionary class struggle. Their greatest goal is the socialisation of all means of production. The means to that goal is the revolutionary class struggle. According to the principles of our brother on the left the antagonism between the working class and the upper classes is unbridgeable. In the German employer they see only their oppressor, in the foreign worker their class comrade ... We say in contrast 'admittedly there exist conflicts of interest, but these antagonisms are bridgeable ... We Germans are a great peoples' family (*Volksfamilie*), we all step up to the great table of the peoples' economy'. The first effort of Germans must therefore be directed to bringing many goods to this table. Then can the distribution of goods for the individual be bettered ... The distribution at the table of our peoples' economy must occur according to the laws of Christian justice and love.

The *Gewerkverein* represented in this speech is virtually indistinguishable from its pre-war guise. Although the use of the word 'brother' hinted at warmer relations between the two unions, the rest of the speech categorically rejects any similarity in principles. Class conflict is rejected and capitalism embraced. The fruits of the economy were, however, to be shared and the best way to ensure the most equable distribution of goods was by the application of Christianity. Moreover, just as in pre-war *Gewerkverein*, conflict was not ruled out. The speaker did admit that sometimes the owners kept an unjustifiably large amount for themselves. Under these circumstances the strike was legitimate once all peaceful means had been exhausted. But in contrast to the *Alter Verband*, the *Gewerkverein*'s 'struggles [were] free from the spirit of the revolutionary class struggle'.[58]

Unsurprisingly, the *Gewerkverein* was even more critical of the *Union* and KPD. Syndicalists and communists were the 'dangerous opponents of the trade union policy'. The *Gewerkverein* contended that the *Union* aimed to overthrow democracy and was not therefore an economic organisation, 'but a revolutionary fighting troop'.[59] For the *Gewerkverein* the left-radical leaders operated under a 'mask of friendship with the workers' for their own political purposes. Yet despite this rejection of materialist discourses, both in social-democratic and communist form, the *Gewerkverein* also had its own version of socialisation. Like Hue, Imbusch saw socialisation as putting the industry in the service of the community. He argued that socialisation did not necessarily mean State Socialism or Communism. Instead he saw it as introducing a moral dimension to capitalism. Private property should not be abolished; rather it should be made subservient to the needs of the community. The 'economic egoism of the individual and the nation' had to be overcome. Socialisation, therefore, represented not merely a change in ownership, but also an ethical transformation. 'The technical socialisation of the enterprise alone is inadequate; it does not tailor the economy to the service of human beings. There must also be a change in spirit. Socialisation is only possible through a corresponding spirit and ethical attitude.'[60]

Imbusch rejected state ownership, believing it did not create the required social spirit, and pointed to the existing state mines to prove it.[61] Instead, the consumers and workers had to be given more rights to balance the power of the employers.[62] One method of increasing the rights of the workers was to introduce small share ownership among the workforce. For Imbusch this meant that the miners would have a direct interest in the health of the industry and would literally share its fortunes. Moreover, it would not result in the loss of initiative he feared would occur under state ownership.[63]

It is clear that the *Gewerkverein* saw itself as steering a middle course between 'extreme capitalism and revolutionary socialism'. Its policy remained 'struggle for a Christian-social community ideal'.[64] Its leaders welcomed the proposals of the Socialisation Committee and they defended the ZAG as a vehicle for improving co-operation between capital and labour,[65] although they, like the *Alter Verband*, maintained that these organs needed to be backed by strong trade unions.[66] However, a change in attitude towards the employers is identifiable after 1923. The *Gewerkverein* regarded the French Occupation as a chance for the owners and miners to show solidarity. At a protest meeting in January 1923 Imbusch admitted that the unions and employers had had their differences, but that they had the German economy, language and culture in common.[67] The MICUM agreement signed by the owners and the French occupying force and the subsequent employers' offensive on hours and wages had a radicalising effect on *Gewerkverein* discourse. Hopes of a new morality in industrial relations dashed, the union became more aggressive in its attitude towards employers from late 1923. Articles in the *Bergknappe* began to complain that the owners had not changed and that they held true to their '*Herr-im-Haus*' philosophy.[68] The 1924 lockout was portrayed as an act of capriciousness, an attempt to repress the workers.[69] By 1925 even the *Gewerkverein* had to admit that the conflict between labour and capital had sharpened. 'Capital, the economy, is everything for the employers, labour only a ware.' Yet despite this more critical attitude the *Gewerkverein* still saw itself as balanced between proletarian and capitalist dictatorships.[70]

As might be suspected the *Alter Verband* maintained a consistently more critical attitude towards the employers. Yet there existed differences of opinion within the social-democratic movement as a whole. The SPD politician, Rudolf Hilferding, for example, struck a more aggressive tone than Hue in his address to the first works' council congress in October 1920. He envisaged socialisation as the creation of industrial parliaments in which the workers, technicians and consumers would be represented.[71] His exclusion of the owners was more forthright. He claimed 'socialism cannot mean a cooperation with the capitalists in any common organisational forms, but can only mean the elimination of the capitalists from production, the elimination of capitalists as owners'.[72] Hilferding portrayed the majority proposal of the Socialisation Commission as *Vollsozialisierung*: full or complete socialisation. 'It demands that the capitalist disappears from the coal industry and that leadership of the industry be transferred to a coal parliament, in which the director of the works, the workers, the officials, the consumers and the representatives of the public are present.'[73]

Unsurprisingly, Hilferding approved of this plan, but regarded it as a minimal demand (*Mindestforderung*). Like Hodges, Hilferding saw the nationalisation of the coal industry as the first step in a wider programme of collectivisation. The German miners, or at least the *Alter Verband* section, were to be pioneers of socialisation, setting an example to other industries. In a passage that illustrates the continuing importance of a working-class identity to the SPD and *Alter Verband*, Hilferding claimed:

> We want to lead the struggle for socialisation of the coal industry, because through it we can hope to once again unite the whole proletariat in a great, united action. This slogan is unifying and this struggle will be decisive. First, because it must and will be a true class struggle, completely undisguised: on the one side the proletariat, on the other the bourgeoisie. We know that the bourgeoisie will employ every force so as not to be defeated; [socialisation] will therefore be treated as a genuine measure of power between the two classes. As such we can only be armed when we enter the struggle united. The struggle is also decisive because we know that, when successfully fought, it places us in a position of power that no one can snatch from us, and from which we can more easily win other concessions. Do not think that we will be content with the socialisation of an individual industry, for then we would not be socialists! But when we have fought the struggle [in this] most important area, further battles will be made easier, not least because the fighting proletariat will have seen what it is capable of when united.[74]

Hilferding repeatedly emphasised in the published speech that nationalisation was a question of power (*Machtfrage*). His discourse reveals how the Social Democrats continued to regard the relationship between employers as one based on conflict rather than cooperation. Just as before the war, the complexities of class formation were simplified down to the proletariat and bourgeoisie, while socialisation of coal was presented as an issue capable of rallying the working class.

In emphasising the opposition of the forces of labour and the forces of capitalism, the *Alter Verband* and SPD ideologues had to tread a fine line. On the one hand, the activists wished to hold on to their identity as opponents of capitalism. Thus Hilferding's pamphlet refers to class struggle between the proletariat and the bourgeoisie. On the other, they needed clearly to differentiate their ideological ground from the Communists. This was especially important given the numerous interpretations of socialisation. Therefore, Hue's assertion that socialisation meant the transfer of industry to the community was not merely aimed at creating a wider body of support outside the union, but was also a conscious attempt to marginalise the communists and syndicalists, which would rob the miners of what freedoms they had, including the right to strike.[75] Moreover, the left-radicals were accused of weakening the trade union movement. In an article entitled 'against the association destroyers', the *Bergarbeiterzeitung* rather optimistically claimed that, had the communists not countered the efforts of the *Alter Verband*, its membership would now have numbered half a million. The paper believed that communists had directed their struggle not primarily against the employers but against the *Alter Verband*. Through its barren agitation the *Union* had succeeded only in driving many away from the unions.[76] Syndicalism too was also presented as a sectional doctrine. 'The socialisation of the coal industry is not a special workers' question. Whoever labels it so maliciously confuses the situation and plays consciously or unconsciously into the hands of syndicalism, which wants to transfer "the mines to the miners".'[77] These attacks on the syndicalists were aimed at establishing a discursive dichotomy based on the sectional interests of one side and the national interests of the other. Overt or implicit, a common feature in *Alter Verband* discourse was the portrayal of the left-wing opposition as un-national. In some respects the social-democratic trade unions had adopted a national discourse.

The discourses of nationalisation/socialisation in south Wales and the Ruhr provide an illustrative example of how identities were differently constructed in the two regions. Nationalisation, as suggested in the previous chapter, provided a focal organisational point for the miners' movement in south Wales. In the SWMF discourse it was mixed into the more inclusive, 'community-fixated' identities that the SWMF and Labour sought to create. The very idea of nationalisation, resisted by the employers, promised by the government, but then snatched away, fitted perfectly within a discourse predicated on a labour versus capital dichotomy. The issue was continually presented in this manner and stretched so that the miners' struggle became symbolic of the fight against capitalism in general. In this manner, even those who favoured socialisation over nationalisation could see it as the important first step. Finally, the experience of the industrial disputes of 1921 and 1926 seemed to dovetail well with the exclusion of the employers from the concept of the community evident in Labour's discourse prior to 1914. This was especially so since trade union and Labour activists were often one and the same. The lockouts and wage cuts made it easy to portray the employers and Government as determined to destroy the mining communities. Nationalisation, it was implied, was their salvation. As notions of community and nationalisation became intricately bound together, Labour was thus able to project itself as the guardian of the former.

In the Ruhr socialisation did not enjoy this wide appeal. The term itself was contested. In different versions it envisaged no role for capital, a limited role for capital or cooperation between labour and capital. Due to these competing visions it could not be constructed into a discourse that claimed to represent the hopes of an entire community. Instead each version was linked to a particular political viewpoint. As Imbusch himself noted it became a *Schlagwort*, used not to unite but to mobilise a certain group of miners, often against another. As a result the very concept reinforced the divided nature of the miners' movement in the Ruhr.

## National identity and class

As noted above, the SPD and *Alter Verband* sought to characterise the syndicalists as un-national. This patriotic element had been evident in SPD discourse in the 1900s, but during the Republic it became more pronounced and fully displayed during elections. Whereas before the war the SPD had presented itself as a working-class party, it now increasingly used the more inclusive term *Volkspartei*. The *Essener Arbeiter Zeitung* (*EAZ*) and *Rheinisch-Wesfälische Arbeiter Zeitung*, for example, rhetorically asked which party was the people's party. Discounting the other parties on the grounds of religion or class, the paper presented the SPD as the only true *Volkspartei*. In fact, during 1919 the SPD attempted to create a *Volkssozialismus* (people's socialism).[78] The claim was linked to the founding of the Republic and the extension of the vote.[79] The SPD made efforts to attract the newly enfranchised female electorate, play down any anti-religious tendencies and broaden its appeal to white-collar occupations. Numerous pieces appeared in the social-democratic press stressing the importance of the female vote. Rather chauvinistically Husemann suggested women needed educating in the democratic process and called on his female comrades to help.[80] It was stressed that the franchise for women was 'the work of the social-democratic party'.[81] Similarly, many of the leaflets produced by the party appealed not solely to the workers but also to the officials (*Beamten*).[82] The *Volksblatt* claimed optimistically that the state officials were now praising the SPD.[83] The trade union pamphlets on socialisation noted the important role the technical staff would continue to play.[84] A priest from Cologne even spoke at an SPD meeting in Essen as part of their 1920 campaign. In a speech that would have perhaps been familiar to voters in south Wales, but was a rarity in the Ruhr, the links between socialism and Christianity were stressed.[85]

This identification with the *Volk* was perhaps natural since much of the SPD pre-war class discourse had been occupied by the USPD and KPD. Before the war the SPD had largely enjoyed hegemonic position in its claim to represent the 'worker'. Admittedly, the Centre had also had a strong working-class element to its discourse, but more often than not this was qualified by an explicit adherence to Christianity. The construction of the SPD as the guardian of the German community and its rights forged during the Revolution also left the party in a position to attack Communist policies perceived as damaging to the *Volk*. In the early years of the Republic this was often expressed as a fear of civil war. Later the focus was on their lack of respect for human rights. Events in Russia provided the SPD with ample propaganda material.[86]

In presenting itself as a *Volkspartei* the SPD was impinging on the discourse used by the other parties, including the Centre. In the 1920s the Centre continued to present itself as the true people's party. This, alongside its Catholicism, remained a characteristic feature of the party. In fact, religiosity remained the defining feature of Centre Party discourse – so much so that some historians have charged the Centre Party with retreating to a narrow clericalism during the Weimar Republic.[87] In fact, many of the party's pre-war concerns disappeared in the 1920s.[88] An examination of the pro-Centre newspapers in the Ruhr seems to bear out this charge. Repeatedly, the editors of *Tremonia, Essener Volkszeitung* and *Westfälische Volkszeitung* produced articles that focused largely on religious issues. A favourite concern was the issue of education. This was given impetus in 1919 when the SPD culture minister, Adolf Hoffmann, pursued a policy of interdenominational schools and abolition of state aid for churches. One contributor described this policy as stripping ethics from the young (*Entsittlichung*).[89] Another called for an alliance between the Catholics and Protestants to ensure the victory of a 'Christian idea of the state', rather than a 'socialist German Republic'.[90] Two years later at a meeting of the Catholic *Volksverein* in Bochum it was argued that religion was the basis of the cultural life of the country and should be the basis on which electors voted.[91] Representatives from Westphalia stressed that the party had been founded as a 'Christian people's party', which sought to base policy on 'principles of Christianity'. Accusations that the party had abandoned religion were strongly refuted.[92]

However, although both the SPD and Centre claimed to be *Volksparteien* there was no blurring in their ideological differences. Indeed there was an intrinsic tension between their attempts to appear as *Volksparteien* and the 'milieu-fixated' nature of their discourse. Both the SPD and the Centre had very precise ideas of what the concept of the *Volk* encompassed and this tended to hold back their attempts to widen their appeal. For the Centre, Christianity and the *Volk* were inseparable and the *Volkspartei* label was more often than not preceded by the delimiting adjective 'Christian'. Meanwhile SPD attempts to extend the party's appeal beyond the traditional working class did not mean that concerns with issues of class lost their vibrancy. From the beginning the SPD defined the Republic in socialist terms. Its defence was placed within the context of long-term social-democratic aims. It was claimed that 'the social-democratic party was always revolutionary in the sense that it strove for the complete transformation of the state to democracy, the economy to socialisation'.[93]

Since the Republic was essentially seen as a socialist creation, the SPD were still able to utilise the older class elements of their discourse. Thus, during the *Nationalversammlung* election in 1919 the *Volksblatt* carried the slogan 'not a capitalist, but a socialist Republic'.[94] Similarly, in 1921 the main question during Prussian *Landtag* elections was 'For or against socialism?'. In differing forms this choice between capitalism and socialism was repeatedly offered to the voters. Even six years after the founding of the Republic, *EAZ* still called the 1925 presidential election a struggle between Republic and monarchy.[95] The paper called on 'republicans' to rally around the SPD candidate Otto Braun and defend Ebert's legacy.[96] Their opponents on the right were still referred to as the *bürgerliche Parteien* (middle-class parties). These parties, it was claimed, included everybody's wishes within their programme, but 'only those of the capitalists' were fulfilled.[97] In SPD propaganda the Centre too was coming increasingly under the influence of heavy industry, while the German People's Party was the party of the managers.[98]

A sense of the difficulty the SPD faced in navigating between these two imperatives is given in an article in the *Volksblatt* on the nature of the class struggle. It noted that, on the one hand, the middle-class parties accused the SPD and free trade unions of preaching class conflict. On the other, the Communists claimed they had abandoned it. 'While the first condemn our class conflict activity and describe it as a disaster for the German people, the other credits it a crime and a betrayal of the German proletariat that we have shed our character of class struggle.' In trying to clarify the situation the article produced a systematic analysis of class struggle that was indistinguishable from the party's pre-war discourse. At a bird's-eye level, it was claimed, the German people appeared united. Beneath the surface, however, there were many divisions, not only in terms of wealth, but also in terms of legal rights and intellectual resources. The knowledge of this inequality ultimately led to the creation of a class consciousness. Class consciousness led to the class will to improve the situation, which was resisted by the property owners. This inevitably led to conflict because:

> when two opposing wills run one against another, a conflict is unavoidable. Class conflict it therefore is also a necessity, a harsh, bitter necessity. It is not some artistic evocation generated by agitation, it is the product of class antagonism and will only end when this opposition is [removed] form the world or when the under classes are lulled or addled to such an extent that class consciousness and class will is dead in them.[99]

This quote illustrates the continued relevance of class and the struggle between classes. Despite its attempt to recruit among the lower middle class the SPD clearly still identified with those of no or little property. The party might have tried to present itself as a *Volkspartei*, but this *Volk* was placed in a similar juxtaposition to the capitalists, domestic and international, that working class had been before the war. It would be wrong to claim that conceptions of the *Volk* and the working class were fully elided. In their references to *Beamten* and

*Angestellten* the SPD was self-consciously extending its electoral embrace. Yet within the discourse the 'Other' clearly remained those who owned the means of production. The issue of socialisation therefore became a totemic demand, proving the SPD continued opposition to capitalism.

But how did the SPD deal with that new 'other', the KPD? The SPD answer to accusations of treachery was to claim that the communists misunderstood the nature of class conflict.

> Like every struggle the class struggle can be fought in different ways. It is an economic and a social, a legal and a political, an intellectual and a cultural struggle. It is also, under certain circumstances a violent struggle. It is unfortunately the ominous error of the left-radicals, that they see the class struggle as merely a conflict through violent methods, while in reality the violent struggle may only be applied when all other means have failed.

Under the Republic, it was argued, the proletariat was in a totally different position than under the *Kaiser*.[100] The abolition of the three-class electoral system, the extended franchise and freedoms of speech and association meant that violence was no longer necessary. Class could be motivated and expressed through the ballot rather than the bullet. The KPD endangered those rights by their advocacy of violence. The SPD did not deny class conflict, but sought to contain it within the political structures of the Republic. Armed actions would break these structures. Similarly, 'wild' socialisation could only serve to destabilise the country. The Republic was poised between the twin threats of capitalism and communism, the SPD its sole guarantor.

In the context of the mining industry this dual threat was coherently expressed in the *Alter Verband* reaction to the 1924 dispute. In the pamphlet the employers were regarded as the aggressors in the dispute. Their demand for longer hours and lower wages and the lockout they initiated to enforce these demands enabled the *Alter Verband* to portray them as union busters. Such was the employers' desire to remove the unions that they were willing to cooperate with the French occupying authorities. As soon as the common front propped up by passive resistance was ended, the employers had negotiated the MICUM agreement without involving the unions.[101] The wage award granted after the four-week lockout was regarded as a victory for the united front of the miners.[102] However, the struggle was not over as some mines had tried to prevent the re-election of works councils by arguing the lockout counted as an interruption in the six months continuous work clause of the works council law, thereby making those locked out ineligible to vote.[103] The picture then is of unscrupulous employers, prepared to use any means to break the ZAG and the unions.

The communists were considered in the second part of the pamphlet. They too were portrayed as out to destroy unionism. The pamphlet alleged that the communists, having failed to realise their general strike strategy, had attempted to present the lockout as just such a strike. This was a 'great danger' since it influenced the press perception of the dispute and alienated the public. In recognition of the weak position of the unions, the *Alter Verband* stressed the importance of gaining public sympathy in any dispute. The combative language of the *Union* and KPD actually weakened the miners. The whole constant denunciation of *Alter Verband* leaders was depicted as destructive.[104] The communists were even accused of inciting followers to murder.[105] In fact, the whole ideological basis of their discourse is thrown into doubt by the authors, who claimed that it had nothing to do with Karl Marx and everything to do with the anarchist Bakunin.[106] Having 'uncovered' the 'emptiness' and 'confused nature' of their discourse, the pamphlet did distinguish between the ordinary members who joined the *Union* and the KPD out of the 'most noble motives', and the 'power hungry clique' which led them.[107] While the employers represented the enemy without, the Communists were the enemy within the miners' movement.

Returning to south Wales, although the particular conception of nationalisation outlined by Hodges was the most widely accepted by the SWMF, and the MFGB as a whole, there existed dissenting voices. Sometimes these voices came from outside the coalfield. For example, John Maclean, the revolutionary Glasgow workers' leader, spoke at Pontypridd to condemn nationalisation and, echoing contemporaneous debates in the Ruhr, support socialisation.[108] In south Wales itself support for socialisation came from some that had been involved in the URC and the drafting of *The Miners' Next Step*. Often this opposition was centred on the belief that the plans for nationalisation did not go far enough. Continuing to follow the themes of industrial unionism and syndicalism that had circulated before, these activists, much like the FAUD and Commission of Nine in the Ruhr, sought to exclude both the influence of capitalism and the state from the industry and establish workers' control.

The most cogent exposition of this argument was put forward by a pamphlet entitled *Industrial Democracy for the Miners: A Plan for the Democratic Control of the Mining Industry*.[109] The boundless confidence of the activists was again expressed in the pamphlet's foreword written by William Ferris Hay, then President of the South Wales Socialist Society, the successor organisation to the URC. 'Nationalisation,' he claimed, 'of the mining and other basic and key industries is obviously bound to come almost immediately.' However, for Hay this did not represent the final step in the miners' struggle against capitalism. Rather it was the penultimate one because nationalisation was 'the last permutation of which Capitalism is capable'. However, under nationalisation the miners needed to 'work out and become thoroughly acquainted with the administrative mechanism, which will ultimately function under Socialisation'.[110]

For Hay nationalisation merely represented an interim or transitional phase. Once the miners were acquainted with the administration any last vestiges of state control could be removed leaving the industry solely in the hand of the miners. In a plan again reminiscent of the Commission of Nine, the pamphlet proposed a hierarchy of committees headed by a Coal Controller responsible for the sale of coal. Authority then devolved downwards through the MFGB Executive, local area executives (i.e. south Wales, Scotland, Northumberland

and Durham, and so forth), colliery lodges, colliery committees and, finally, pit committees.[111]

While the supporters of socialisation shared much the same criticisms of capitalism as the 'nationalisers', they were concerned that nationalisation would mean 'officialisation'. While Hodges believed that disputes would disappear in the industry once the profit motive was removed, Hay believed that control would still remain a contentious issue. 'It would afford no consolation to the miner to know that he was exploited to relieve national taxation instead of for private profit.'[112] Only when the miner had the greatest possible control over his work would that cease to cause disputes.[113] The solution to this problem was to turn control over to the MFGB, with whom the government-appointed Coal Controller would place production orders. Hay was critical of the MFGB, believing it to be a crude instrument for the exercise of democratic control. However, Hay thought that the MFGB provided a 'skeleton structure' which could provide the basis of democratic control. 'It is co-extensive with industry; it has the forms and machinery – albeit imperfect – for democratic control, and with the alterations and improvements we suggest, could well become the steward to administer the nation's coal resources.'[114]

Although the plan for socialisation of the British industry bore many similarities to proposals in the Ruhr there were three fundamental and interrelated differences between the Welsh and German 'socialisers'. First, just as the URC members saw no need to establish a union to rival the SWMF before the war, so too did the 'advanced' men of the post-war period see reform of the MFGB from the inside as their best option. This was in contrast to the syndicalists in the pre- and post-war Ruhr, who saw counter-organisations as the only way to advance their cause. A more flexible attitude from both the advocates of nationalisation and supporters of socialisation, on the other, contributed to the containment of potential tense relations. Just as before the war, both sides were willing to tolerate a plurality of approaches within the same organisation, rather than seeking to impose harmony by excluding rivals.

Second, the similarities in their discursive identity conditioned the containment of disputes and the flexible attitude of activists. This is not to suggest that disagreements between SWMF activists were without acrimony. This was especially evident during the 'Datum Line' strike of 1920.[115] Radical leaders, such as S. O. Davies, A. J. Cook, Ablett and George Barker supported a stronger stance by the MFGB, while Brace, Tom Richards and Hartshorn sought a more conciliatory line.[116] Brace was critical of the highhanded attitude of 'direct actionists'. He believing they adopted 'the airs of the superior person and ... act as if their higher mental equipment conferred upon them the right and authority to order masses of their fellow-workers to cease work without their having been consulted'.[117]

Such was Hartshorn's anger at the extremists that he resigned his post on the SWMF Executive after the strike, claiming 'an atmosphere of hostility and suspicion [had been] created against leaders who believe in steady revolutionary reform'. This atmosphere was, according to Hartshorn, the work of 'dangerous revolutionaries'.[118] In fact, despite his early attendance at URC meetings Hartshorn had grown increasingly critical of the Committee's radicalism. This is illustrated in his response to the URC's planned down-tools policy in 1918. He disparagingly termed the Committee 'self-elected and self-appointed' and argued that few had stood in ballot against SWMF officials and where they had they had been defeated. Hartshorn believed the URC to be a danger to democracy since 'instead of winning the right to speak in the name of the rank and file, [they] proceed by methods which are the negation of democratic control'.[119] Bitter recriminations were also made after the 1921 lockout. Hodges was critical of what he termed the 'Ablett type', who was 'responsible for prolonging the stoppage'. Ablett's response was to portray Hodges as a man 'pitchforked into a position of responsibility before experience had prepared him for it'.[120] Moreover, these struggles over the issue of tactics stretched down to the lower levels of the miners' organisation. The minutes of the Ferndale lodge, for example, bear testimony to the struggle between Noah Tromans and Charlie Gibbons on the issue of political as opposed to industrial action.[121]

However, these disputes were often representative of personal rivalries and were generally of short duration. Hartshorn, for example, eventually returned to the SWMF Executive in 1922. Ultimately, both the leading supporters of nationalisation and socialisation in the SWMF propagated an identity based on the dichotomy of miners versus employers. This common opposition to capitalism provided an effective rallying point in a way it could not in the Ruhr, since the *Gewerkverein* was firmly committed to private property. The miners' leaders in south Wales also recognised the importance of propaganda in maintaining this identity. Following the defeat in 1921 John Thomas, miners' agent for the anthracite district, argued that 'propaganda must be extensive and intensive. Lodge meetings must be turned into lecture rooms and classes, centres of discussion and planning, about the next phases of the struggle, to win the control of their Industry for the workers.' To facilitate the dissemination of the union's discourse, Thomas advocated the production of a newspaper.[122]

Undoubtedly Thomas would have been pleased with the production of *The Colliery Workers' Magazine* in 1923. Its pages consistently reinforced the basic miners versus employers dichotomy of SWMF discourse. It would be repetitive to examine every permutation of this discursive practice, but perhaps Oliver Harris can be regarded as typical of the general view when he described the industry as being 'in the claws of Capitalism, with its wasteful, inefficient and unscientific administration'.[123] The immutable nature of the conflict was constantly stressed. Ironically, W. H. Mainwaring's argument, while rejecting the notion that class itself was a formation of ideas, also strengthened the class nature of the union's discourse. Rather class was formed by economics, by the sale of one man's productive labour to another. 'They [the workers] stand on one side as sellers of their capacity to labour. Opposed to them are the would-be buyers of this labour power'.[124]

Yet while the *Colliery Workers' Magazine* was unequivocal in its opposition to private ownership, in other respects the paper provided a surprisingly open forum for debate. In fact the magazine was a prime example of how the SWMF allowed pluralist debate to be expressed within its organisation. All shades of the nationalisation/socialisation debate were represented in its pages. Seemingly the only proviso was that the contributors were opposed to private enterprise. This openness also extended to other areas. One example was the open debate on affiliation to the Red International of Labour Unions (RILU).[125] Characteristically, the *Colliery Workers' Magazine* presented both sides of the debate in a single issue. John Thomas presented the case for affiliation. He described the International Federation of Trade Unions as a bulwark of capitalism, but by joining the RILU the SWMF would be moving a step closer towards abolishing capitalism as well as living up to its 'pioneer' role. Arthur Jenkins, miners' agent for the Eastern Valleys District, presented the counter argument. The crux of his argument centred on the split that affiliation would create. This could only weaken the labour movement.[126] The contours of the debate are here less important than their presentation. Printed side-by-side the two pieces demonstrated the magazine's commitment to providing an opportunity for all to voice their opinion. Thus the magazine acted as an important ideological safety valve. The differences between 'nationalisers' and 'socialisers' were aired, rather than suppressed. This prevented discontent from festering and helped maintain the organisational unity of the SWMF.

In the Ruhr, on the other hand, the issue of affiliation to the RILU was far more charged. Again, the more 'pillarised' civil society in the Ruhr precluded such wide-ranging debate on the issue. Instead, affiliation or non-affiliation became a badge of membership to a particular political ideology. For the communist-syndicalist unions membership of the RILU symbolised the truly revolutionary nature of their organisations. At the founding congress of the *Union* affiliation was seen as part and parcel of struggle against the 'dictatorship of bureaucracy' which dominated the social-democratic unions.[127] In October 1922 the *Union* published a letter asking the *Alter Verband* to join it in common action. The *Alter Verband* rejected the request, citing the *Union*'s membership of the RILU as a reason. As far as the *Alter Verband* was concerned the RILU was dedicated to its destruction.[128] Thus, rather than being a tactical question, the affiliation issue was seen as affecting the very existence of miners' unions. Under these circumstances the 'fair' debate style of discourse that characterised the contributions to the *Colliery Workers' Magazine* was a rarity rather than the norm in the trade union press of the Ruhr.

While rivals within the SWMF might castigate each other as foolish or naive, and use derogatory terms such as the 'Ablett-type' or 'Hodgean Mind',[129] in general neither side sought to label their opponents as the deliberate enemies of trade unionism or of the working class itself. It is noticeable, for example, that, despite his criticisms of the 'direct actionists', Brace did not reject the idea of the strike as a weapon, merely its timing. Similarly, although he believed *The Plan for the Democratic Control of the Mining Industry* misguided and an 'intellectual conceit', he admired the enthusiasm of its authors. He was in no doubt that this enthusiasm was of use as long as it was 'controlled, organised and founded on loyalty to the movement'.[130] This respect for the effort underlying these ideas, but rejection of their plans, is well illustrated by Hodges's pamphlet *Workers' Control in the Coal Mining Industry.* While the pamphlet contains much of the same material present in Hodges's book, it also touches on Hay's plan for democratic control. Hodges rejected the scheme, but he recognised it was 'the work of extremely thoughtful men'.[131] Although he later went on to describe the 'futility' of syndicalism and its 'anti-social' characteristics,[132] he accepts the psychological background of antagonism that lies behind this plan as the defining characteristic of the industry. Hodges then, while disagreeing with the blueprint for control, validates the syndicalists' concerns and even pays them a compliment. It is implied that the syndicalist plan is one among many, but, given due consideration, is not the best one. Its authors are not identified as enemies, but are included within those trying to work for the betterment of the miners' position (even if they had got it wrong). Nowhere in the document is there the hyperbole that, more often than not, characterises dialogue between the unions in the Ruhr. Even in his criticism of the 'Ablett-type', Hodges does not claim that the strike was unjustified or blame it on those individuals. He restrains his criticism to their role in prolonging the conflict. The whole issue has the feeling of a disagreement over tactics, rather than fundamentals. Opponents were not regarded as constructing an alternate identity of the miner in their rhetoric. Indeed, they did not create alternatives to the prevailing labour versus capital model. More often than not rivals were seen as merely misguided, while the sense of frustration and dislike of capitalism underlying their work was met with understanding.

The shared opposition to capitalism of the 'Ablett-types' and 'Hodgean minds' fed into the third difference between trade unionism in south Wales and the Ruhr. For although they differed over whether nationalisation represented an end goal or a stepping stone to socialisation, Welsh trade union activists were, in the final analysis, agreed on the desirability of its immediate introduction. In the Ruhr the various plans for socialisation were mutually exclusive. As the AAU pamphlet claimed it was a question of either-or. In south Wales, on the other hand, the different plans were not mutually exclusive. The more pragmatic culture of the SWMF meant that rather than a question of either-or, it was a question of nationalisation now and (for some) socialisation later. The employers' determination to retain the industry in private hands and the government's 'betrayal' provided both 'nationalisers' and 'socialisers' with a common platform.

The unequivocal rejection of nationalisation by employers and the government meant that the SWMF activists could indulge in a clear, unambiguous 'othering' of their opponents.[133] In other words, the battle lines between those opposed to and those in favour could be clearly drawn. The TUC publicity campaign, and consequent flood of pro- and contra- literature, heightened this employer/government versus miners dichotomy. The MFGB, for example, produced a series of twelve pamphlets to support their case. Many drew on evidence submitted to the Sankey Commission to highlight the supposed wastefulness of capitalism. Others struck a humanitarian and moral note, carrying headings such as 'Every day four miners are killed!' and 'Nationalise the mines and save the lives of men!'[134] Still others, in keeping with the 'Mines for the Nation' slogan, sought to underline the benefits of nationalisation to the wider community. Pamphlet ten targeted the consumer, while number nine appealed directly to the middle class by stressing how the official would also benefit.[135] A similar line was adopted by the pamphlets *Facts from the Coal Commission* and *Further Facts from the Coal Commission*. The latter especially made emotive use of notions of betrayal. The editor and historian of the miners' movement, Robin Page Arnot, cited Hartshorn's reaction to sum up this feeling: 'Was it a huge game of bluff? Was it never intended that if the reports favoured nationalisation we were to get it? [The miners] will say "we have been deceived, betrayed, duped".'[136]

### Militaristic discourse and 'enemies'

Perhaps the most emotive claim used by the unions was that of profiteering. In a reworking of Government wartime propaganda, pamphlet two carried the heading 'What did you do in the Great War, Mr. Coalowner?' It claimed that the owners' profits had risen by 270 per cent.[137] Naturally, nationalisation meant an end to this profiteering.[138] Instead, the industry would be put at the service of the nation, just as it had been during the war. In this context, the constant reminders of the dangerous nature of minework can, therefore, be linked to notions of sacrifice and selflessness that arose from the war. Hartshorn drew a direct comparison between the conflict and the industry in his evidence before the Sankey Commission. 'In the mining industry the casualties are more like those of the battlefield than anything else. The only difference between the soldier and the miner is that the miner can never ask for an armistice. He cannot even treat for terms of surrender. The casualties go on every day.'[139]

War metaphors were commonplace during the great strikes. A. J. Cook claimed at a meeting in Gwaun-cae-Gurwen that the 1921 lockout was 'a war of attrition and starvation that the Government were waging to conquer … Our battle is the battle of the workers of the world'.[140] In 1926 Cook described the strike as an 'industrial war' and claimed the MFGB executive 'is now a council of war'.[141] On one level such war imagery was used to construct feelings of solidarity among the men. However, it also accentuated the notion that industrial relations were essentially based on conflict and helped further bury more collaborationist trends.

Another problem for collaborationist thinking was the attempt to depict the industrial disputes of 1920, 1921 and 1926 as not merely an issue affecting the miners, but as symptomatic of the essential conflict between labour and capital. It was in this vein that S. O. Davies claimed in 1921 that:

> he could not comprehend how any worker, whatever the industry he found employment in, being able to circumscribe the crisis as being merely a conflict between the mineowners and the miners of this country … The organised forces of capital were momentarily confronting the organised ranks of the miners, but the

> defeat of the mines, if it could be consummated, would never satisfy organised capital. The other workers were asked to come into the fight not because the mining forces were disintegrated, but because the fight was one of organised Capital versus the whole of organised Labour.[142]

Such monolithic characterisation of labour and capital made it difficult to find a common basis for negotiation. Conflict was seen as the existential core of the relationship between the two sides. Within this discourse any agreement would ultimately seem more like a temporary truce in an inevitable conflict than a step toward lasting cooperation.

The miners' leaders in south Wales were not the only ones to feed this discursive construct in speeches and publications. The employers too drew analogies between the industrial disputes and the Great War. For example, in 1919 the *Colliery Guardian* commented in an article provocatively entitled 'Take the Gloves Off!' that:

> the question has long since ceased to be a coalowners' question; they merely occupied the front trench. It is now for all the interests, small and great, to rally to the standard in the fight against unprincipled disorder, which will surely, if this picket is overrun, spread to every branch of the national life. Every tradesman and householder must be brought to see that the portent of these preliminary skirmishes is that England will change hands, not merely the land but everything on it and under it, and that the ultimate possession, after a brief and disastrous hegemony, will be an oligarchy embracing many of those whose earnest endeavour has been to surrender the country to its enemies.[143]

Ross McKibbin has argued that the war militarised the mentality of much of the middle class,[144] thereby making the imagery of trench warfare and heroic defence of the nation highly evocative. Like the union activists, the employers were intent on widening the terms of the conflict beyond the industry. The miners, or more often their leaders, were placed beyond the pale, described as un-national and equated implicitly with Britain's wartime enemies. This language was evident both in the external and internal organisational discourse of the employers' organisations. Evan Williams, head of the MSWCOA, told other Welsh owners that it was 'in the interests of the Country [that] Nationalisation should be opposed'.[145] In this portrayal the employers became protectors of the nation, while depictions of the miners' movement were fed by fears of Bolshevism. At the annual general meetings of the Powell Duffryn Steam Coal Company, the Chairman, John Shaw, repeatedly extorted the shareholders to canvas their MPs to reject nationalisation.[146] Following the 1921 dispute he warned the shareholders not to become complacent. 'Do not think for a moment, ladies and gentlemen, that we are out of the wood yet. The Communists and Bolshevists are just as lively as they ever were.' To emphasise the ever-present nature of this threat he pointed to the engineering strike and the Irish troubles. He could only hope that the men had begun to realise 'that their leaders have not been leading them aright'.[147] Like the SWMF, the MSWCOA and other employers' organisations also represented the conflict as having wider implications for the nation.

The south Wales owners also had an outlet for their discourse in the employer-friendly *Western Mail.* In the course of 1919 the paper reproduced the proceedings of meetings of various coal companies which all ultimately ended with resolutions against nationalisation.[148] It also expressed the view that the strike wave that afflicted Britain was due to the influence of Central Labour College classes, which spread Bolshevik and syndicalist ideas.[149] A recent suggestion that the miners generally disliked the *Western Mail* has been countered by the assertion that it was available in all miners' libraries.[150] Whatever the miners' attitude to the paper generally, trade union activists at least regarded it as an employers' paper, so much so that it was used to reinforce their own arguments. Cook, as part of the RILU campaign stated, 'take note, Welsh miners, the coalowners' paper the *Western Mail,* warns you against the Red International – that should be sufficient for you to decide'.[151] Meanwhile, the *Labour Voice* was certain that the *Western Mail* had been invaluable in securing the election of George Barker at the 1921 Abertillery by-election. 'An invaluable aid to Mr Barker was the opposition of the Western Mail. The miners saw a journal, which is avowedly and openly conducted in the interests of the coalowners vigorously attacking their leader, and they felt it their moral duty to come to his aid.' The *Labour Voice* went on to convincingly claim that the miners may criticise their own leaders, but like a family quarrel they did not want anyone from outside to interfere.[152]

War imagery was even more widespread in the Weimar Republic and was given special resonance by the high degree of political violence. The experience of the Front had a destabilising effect on the Republic's political culture.[153] Battles between the Ruhr Red Army, Free Corps and police provided martyrs and legends for both sides.[154] Generally, the Communists made most use of images of war and martyrdom. The memories of Karl Liebknecht and Rosa Luxemburg, for example, were transformed into alternative, communist celebrations.[155] A Hamborn Communist constructed an altar to Liebknecht and Luxemburg.[156] More locally, disasters such as the *Minister Stein* explosion, which killed 136 miners, provided material for propaganda attacking both the employers and the other unions. Like the SWMF, the Ruhr Communists sought to widen the struggle to encompass the whole of the working class.[157] The *Alter Verband,* however, was critical of using these tragedies for such purposes, although it also took the opportunity to attack the profit interest in the mines.[158] Eric Weitz has noted the KPD's need for enemies. He argues that in the vitriolic attacks the Communists made on their enemies the party was tapping into real working-class experiences of humiliation by officials or harassment by the police.[159] Furthermore, Weitz has highlighted the militaristic ethos which pervaded communist literature. For example, a common trope of cartoons and photographs were identifiably working-class men bearing arms in defence of the Soviet Union.[160]

However, although SWMF activists may have largely accepted internal disagreement and closed ranks when faced with criticism from without, this toleration was not extended to those who broke away from the union.

Opprobrium was heaped on the South Wales and Monmouthshire Mechanical and Surface Workers' Union and a vigorous campaign waged against it. A leaflet castigated the Mechanical Union as a 'parasite organisation', before claiming that 'craft unionism breeds disloyalty, disunity, disorganisation, and chaos, and we mean to obstruct and sabotage all attempts to steer for such a cause'. They were depicted as hopelessly inadequate in the face of monopoly capitalism, and 'about as evenly matched as the "Bow and Arrow" would be against the modern "Machine Gun"'.

> Let us point out to them that the fact [that] their belonging to a 'Scab Organisation' will make no difference in our Strike Policy ... Whatever any combination of Blacklegs may decide to do in the interests of their masters, the Federation will have something to say on the question as to whether those men shall be allowed to work or not. Let them not build for themselves a 'Fool Paradise' ... Let the Mineworkers of Mardy realise that the Federation is not dead. It was only temporarily stunned by the blows of its enemies. The Assassin Knife of Craft Unionism, now being used in the attempt to finish off the Giant, will also fail to accomplish the deed. Staggered though it was by the blow, and the reverse of the late Lock-out, the Federation still lives, and its power and vigour is returning. Before long, by loyal service from all workers in the Industry, we shall yet attain one Union for the Mining Industry, and thus build up the Machinery which will make for the emancipation of our Class.[161]

The use of the terms 'scab' and 'blackleg' to 'other' the splinter union had roots stretching back into the pre-war period. Furthermore, the idea of dealing a deathblow to sectional unions had been raised soon after the formation of the SWMF. However, the discourse employed in the pre-war period was certainly not so vitriolic. Evident in the criticisms of craft unions and reference to one union are the echoes of syndicalism. Rather new also was the notion of betrayal. Such language was perhaps unsurprising in Mardy, that 'Little Moscow' of the coalfield. But criticisms and even the interruption of Mechanical Union meetings also occurred elsewhere.[162] The Mardy leaflet may represent an extreme form of SWMF discourse regarding the Mechanical Union, but it also demonstrates both the more aggressive tone the organisation had adopted since the war and the odium that any attempt to form, or reform, alternative union structures would attract.

The Labour Party in Wales also maintained a flexible and broad-based discourse. The most significant change in the immediate post-war era was the breakdown of the progressive alliance between the Labour and Liberal parties.[163] The increasing emphasis on the working class identified in the immediate pre-war discourse was accentuated in the 1920s. Paralleling the weakening of the class collaborative trend in the SWMF discourse, so the Labour Party's juxtaposition of the working class against the employers became even more pronounced. Unlike the SPD, which had adopted and adapted the idea of the *Volk* through its identification with the Republic, the Labour Party in south Wales increasingly moved away from the *gwerin*, supplanting it instead with the Welsh working class.[164]

Elements of that older discourse did survive. For example, an article in the liberal journal, *The Welsh Outlook*, in 1919 seemed to commit the Labour Party to greater autonomy for Wales.[165] Election material and political speeches also contained traces of old Liberal concerns. One such example was T. I. Mardy Jones. In the 1922 election he defended his Pontypridd seat against Sir Rhys Williams. Williams was castigated not only as 'a brewer, [but also] a landlord, and a coalowner'. Here Labour sought to combine the older concerns of temperance and agricultural grievances with the problems of the modern proletariat. More typical of Labour's discourse at the time was the elision of the Liberal and the Conservative parties. Both were represented as the political representatives of capitalism.[166] Similarly, in Ammanford James Griffiths accused the Liberal Coalitionists of betraying 'those principles of Free Trade and Welsh Disestablishment which they had hitherto championed'.[167] A week later it was claimed 'the Liberal Party and the Tories had united together, and they were opposing Labour, and there was now a fight between Capital and Labour'.[168] A potted biography of Griffiths stressed his working-class origins.[169] At a later election the attitude of Griffiths supporters were even less ambivalent. The demand was 'let a Labour majority dictate to the mineowners'.[170] There is little sense of a cooperative relationship in these statements. Interestingly, however, the latter was attributed to one Revd David Richards. Another preacher, James Griffin, also supported Griffiths' candidature. The mixed reaction of Nonconformity towards the labour movement still characterised the political culture of south Wales and undoubtedly contributed to Labour's hegemony despite the increasing secularised culture of the valleys. Indeed, there was a high proportion of lay preachers and deacons among Labour's Welsh MPs, which suggests that Nonconformity had now become more of an asset for Labour than Liberalism.[171]

Unsurprisingly, the CPGB shared this elision of Liberalism and Conservatism. Significantly, unlike the KPD in Germany, the CPGB more often than not supported Labour candidates rather than challenging them. Relations between Labour and the CPGB could be tense. The *Cymmer Searchlight* complained,

> how can the National E.C. of the Labour Party instruct Local Labour Parties to expel Communist[s] with the threat that they will expel them from the Labour Party if they will not carry into effect their decision, when at the same time prominent members of the E.C. are feasting, drinking and holiday making with the Boss Class. Comrades we ask you, who are the workers friends? The Communist now arrested or these so called Labour leaders?[172]

Yet despite these criticisms the pit paper encouraged its readers to vote for David Thomas, the Labour candidate for the Cymmer ward in the local election. It identified a class bias in the local elections, the Conservatives and the Liberals having sunk their differences, just as at parliamentary level, to vote against Labour.

Part of Labour's replacement of the *gwerin* with the Welsh working class involved claiming the Liberals' mantle of progressivism. In fact, one criticism focused on the idea that Liberal policies had become out-dated. In 1924 the Liberals produced a pamphlet entitled *Coal and Power* which set out its policies regarding the coal industry. The review of the pamphlet by D. J. Williams in the *Colliery Workers' Magazine* can be regarded as characteristic of the general view of the Liberal discourse by south Wales Labour activists. The pamphlet itself argued that industry and commerce were the basis of prosperity. Royalties, however, were the chief evil as this represented unearned wealth and made the industry less competitive. Williams saw this pamphlet as an attempt to revive enthusiasm for the old Liberal shibboleth of anti-'landlordism'. He believed the historical platform of liberalism, based on free trade and land reform, was largely irrelevant to contemporary politics. Liberals were guilty of blaming the landowners alone for the troubles of the industry, while ignoring the pernicious effects of private enterprise itself. 'The miner will be consulted, but not trusted. There is, of course, the usual platitudinous wail about miserable housing conditions in the mining villages, and the blame for this is placed everywhere but at the door of private enterprise.'[173]

The notion that Liberal discourse largely remained based upon the same ideas that it had been before the war is also borne out once the electoral language of the candidates is examined. For example, the Liberal candidate for Pontypool in 1924, Major Beaumont Thomas, followed the well-established 'friend of the workers' track. Admitting he was an employer, he added that 'he happened to count among his best friends wage-earners and he said that no employer was worth his salt if he could not say he could also represent his employees'.[174] Others, like Captain R. T. Evans at Llanelli, emphasised their working-class origins. In 1923 he claimed that 'his father was a collier, and he was one of six kids, so that he knew something of the atmosphere of a collier's home'.[175] In a more direct allusion to the tradition of Lib-Labism, David Rees supported the Labourism of Mabon. 'I am not here to fight Labour as such ... When Mabon stood for the Rhondda I voted for him. Therefore, I am not unfriendly to Labour. Why do I fight Mardy Jones? Because he is for Socialism, and I am against it.'[176] Liberals were not the only ones to claim working-class credentials. David Evans, the Unionist candidate for Pontypridd in 1924, 'claimed he had a greater right than his opponent to describe himself as a working man candidate. On the night of his adoption he changed into his working clothes and was at work that night. If he were defeated in the fight, he would not be afraid to return to the collieries again.'[177]

The 'people' and Welsh identity also remained important aspects of Liberal discourse. Many Liberals lamented the passing of the progressive alliance. Clark Williams, Captain Evans' predecessor as Liberal candidate for Llanelli, explicitly tried to rally Liberals around 'the Old Flag'. 'He referred to the struggle between the Liberal and the Labour Party. He wondered how it had come about, for time was that Liberalism stood for no class, but for the welfare of the people as a whole.'[178] As might be supposed, in the more rural constituencies Welshness also remained important. Sir Ellis J. Griffiths at Carmarthenshire, for example, admitted that he was 'not a Carmarthenshire boy, but he was a thorough

Welshman who loved the language and ideals of Wales and he had yet to learn that a Welshman was a foreigner in the county of Carmarthenshire'.[179] Yet national identity was also evident in Liberal discourse in the more industrial, western constituencies. At Caerphilly S. R. Jenkins 'emphasised that he was a son of Wales, was one of the people, for the people, and had striven through life for Liberalism as a creed for the advancement of human happiness'.[180] David Rees was even more impassioned. His election address stated:

> He is a Welsh-speaking Welshman, and his feelings are strongly nationalistic. He even hopes to live to see the day when there will be a Welsh Parliament at Cardiff that will deal with distinctly Welsh questions. He is a strong supporter of the Welsh language, and the 'language of Paradise' is the language he uses in this own home. He believes that Liberalism will give to the worker all and more than Socialism would give him, and that personal incentive would remain, whereas under Socialism it would vanish.[181]

The continuities between the features of this discourse and pre-war Liberalism are plain to see. Yet despite their claim to represent the 'people' it is clear that Liberal discourse had lost much of its ability to encompass the widest cross section of the community possible. The erosion of the *gwerin* ideal that was already evident in the 1900s was accelerated after the war. Several interrelated factors contributed to this decline. In one respect the Liberals can be regarded as a victim of their own success. The disestablishment of the Church robbed religious motifs of their immediacy. It is noticeable that in all the examples given above the pre-war Nonconformity/Welsh–Anglican/English discursive dichotomy does not appear. Moreover, as shown, Labour had successfully adopted much of Liberalism's progressive rhetoric. Finally, a discourse based on the cooperation of labour and capital seemed hardly to correspond with the reality of bitter, months-long industrial disputes.

If Liberalism was a stagnating force in the mining valleys during the early to mid-1920s, there was another alternative discourse to the Labour-SWMF hegemony. This discourse had its roots in Conservatism. Like pre-war Conservatism, much of its appeal focused on a distinctly British identity. During the 1918 'Coupon' election there was some overlap with Coalition Liberals, as the Government sought to capitalise on its success in conducting the war. Much of Coalition electioneering in 1918 was shot through with an anti-German, 'Hang the Kaiser' mentality. Captain Seaton, the Conservative candidate for Pontypridd, based his campaign on this attitude.[182] Later Seaton withdrew and threw his support behind the Coalition Liberal T. Arthur Lewis, who also campaigned on the basis of victory over Germany. He contrasted the victorious Coalition to the ILPers, implying that the latter would have led the country to defeat. 'The Coalition Government destroyed Prussian Militarism which would have been victorious if the ILPers and conscientious objectors had had their way and it is the Coalition Government that should have the carrying out of the peace settlement.'[183] C. B. Stanton, the former working-class firebrand, also laid heavy emphasis on 'Britishness' at Aberdare. For him the war had apparently represented an epiphany as 'he discovered for the first time that he was British'.

He pledged to conduct himself in parliament 'by the memory of a million dead and a million maimed and carry on as a Britisher'.[184] But while such a discourse helped him to victory in 1918, by 1922 much of the 'transitory war mood' that underpinned his earlier victories (and those of coalition candidates) had evaporated. Instead, the class identity expounded by Labour ultimately proved stronger than appeals based on national duty.[185]

There were also conservative currents within the SWMF. As already noted, several lodges initially rejected nationalisation. Mike Lieven has used the example of the Ynysybwl lodge to highlight the plurality of political beliefs in the mining valleys. Here the local Tories were able briefly to gain control of the lodge committee.[186] D. B. Jones and his Mechanical Workers had been, however briefly, members of the SWMF. Undoubtedly, the shattering effects of the 1926 dispute weakened the SWMF position, sapping its financial and human resources, but the activists who formed the breakaway industrial union in 1926 did not emerge from the ranks of the unorganised. George Spencer had been an official in the Nottinghamshire Miners' Association.[187] In south Wales, it was former SWMF members led by William Gregory who broke away to form the organisation that eventually became the SWMIU.[188] Its rejection of politics and industrial action can be regarded as a reassertion of a more collaborative and conservative strain of discourse. SWMF discourse, hegemonic for so long by virtue of its flexibility, had lost its ability by the end of 1926 to flatten out the differences between the miners. It remained dominant, but, temporarily at least, the process of homogenisation that characterised its discourse from 1898 was broken. However, as a corollary the loss of industrial strength caused a shift to the political, reinforcing the mining valleys as a Labour stronghold.

## Summary

In the early 1920s both continuities and changes are identifiable in the discursive practices of the miners' trade unions and the parties. For the SWMF the war had been a confidence-building experience and passed the initiative to the more combative discourse that existed before the war. The more collaborative discourse within the SWMF was submerged, but not banished, by a more aggressive tone that saw the differences between capital and labour as irrevocable. The solution to the conflict could be found in nationalisation, which entailed the abolition of capitalism. Nationalisation provided the perfect platform for this language. The employers' resistance to state control meant that it could be easily held up as an example of the essential conflict of interest between labour and capital. Leaders and activist made the issue symbolic of this wider struggle. Admittedly, some individuals sought a more radical solution in socialisation, but even they believed that nationalisation was a necessary first step. Nationalisation, therefore, provided a common basis for action. Moreover, despite the more aggressive discourse of the SWMF, it remained pragmatic. Its toleration of different ideas was represented by the editorial policy of the *Colliery Workers' Magazine*. Issues such as nationalisation, socialisation and affiliation to the RILU were freely debated. However, the debilitating effects of the 1921 and 1926 lockout ultimately sapped this confidence and the flexibility of the unions' discourse. The more collaborative trend was eventually able to reassert itself in the form of the SWMIU.

The Labour Party's commitment to nationalisation provided a political parallel to the industrial sphere. In south Wales Labour was not totally successful in 1918, but by the mid-1920s had gained control of the mining constituencies. But support for nationalisation was not Labour's only prerequisite for success in south Wales. The party benefited by taking on some features of Liberal discourse. Just as the collaborative strain in SWMF discourse declined in the 1920s, so too did the cooperation between labour and capital advocated by the Liberals become equally unconvincing in the light of the industrial disputes of the 1920s. On the other hand, Labour's political language, which excluded the industrialists from the community in the same way that Liberalism had excluded the landowners, fitted well with the atmosphere of industrial conflict. Notions of the community and the working class were increasingly elided. While Liberalism could only offer a modified version of its *gwerin*–landowner dichotomy through the abolition of royalties, Labour's support for nationalisation allowed it to project itself as the guardian of the proletarian community. By combining this with the acceptance of some Liberal concerns, Labour was able to usurp the mantle of the progressive party. Finally, although the Communists remained among Labour's fiercest critics, they regarded themselves more as a watchdog than a rival organisation and usually threw what influence they had behind Labour. In another example of the greater 'fluidity' of civil society in south Wales, Communists and Labour generally regarded themselves as members of the same 'family' for much of this period and dual membership of the CPGB and other Labour organisations was tolerated. Thus, unlike the Ruhr, where divisions between Communists and Social Democrats were sharply drawn, relations between Communists and Labour activists in south Wales were highly fluid before 1927.

In the Weimar Republic the fragmentation identified during the *Kaiserreich* was accentuated by the war. The SPD was unable to contain internal tensions generated by the war and the split in the social-democratic milieux was to prove crucial. Whereas Labour was able to largely monopolise a working-class discourse in south Wales and in Britain generally, in the Weimar Republic the SPD found itself challenged by the USPD and later the KPD. The fluid relations evident in south Wales and other British industrial areas were impossible between the KPD and SPD as each simply rejected the other's claim to represent the working class as illegitimate. Similarly, although the *Alter Verband* had grown closer to the *Gewerkverein*, it had also splintered on the left and now had to deal with the various syndicalist and communist unions. In some respects the KPD and Syndicalists represented an assertion, or reassertion, of a discourse that already existed in the labour movement before the First World War. The Commission of Nine, for example, was aware of its social-democratic roots, while the Syndicalists drew on the even older tradition of pithead meetings. The

SPD found that to some extent its claim to be the workers' party had been hijacked. It sought to adapt to this challenge and the changed political environment by embracing a more inclusive discourse represented by the *Volk*. However, the *Volk* used by the SPD was intimately linked to the socialist Republic. This differentiated it from the Centre, which clung to a political language based on Christian belief. For both parties there were tensions between their attempts to reach out to a broader electorate and their older 'milieux-fixated' discourses. The discourses of both the Centre and the SPD underwent modification, which saw new force given to some older elements (e.g. patriotism for the SPD) or the introduction of new ideas (e.g. republicanism for the Centre). But essentially both remained predicated on the overarching themes which had dominated their pre-war identities: class and religion.

Finally, while nationalisation was a broadly understood and supported concept in south Wales and the other British mining communities, in the Ruhr socialisation was a confused and shifting idea. Each group had its own version of socialisation and each was ultimately incompatible. It was used to mobilise a certain constituency, not the miners as a whole. The organisations' intolerance of dissent meant that to integrate their different versions of socialisation meant risking their existing support among the miners. The affiliation debate around the RILU provides an example of how dual membership was often not tolerated. The doctrinaire understanding of such issues precluded any enduring attempt at cooperation and perpetuated the fragmented nature of the labour movement.

## Notes

1 *CNWWGA*, 7 December 1923.
2 Stenbock-Fermor, *Meine Erlebnisse als Bergarbeiter*, p. 77.
3 R. McKibbin, 'Classes and cultures: A postscript', *Mitteilungsblatt* 27 (2003), 153–66.
4 See Geary, 'Myth of the radical miner'.
5 Neunerkommission für die Vorbereitung der Sozialisierung des Bergbaues, *Die Sozialisierung des Bergbaues und der Generalstreik im rheinisch-westfälischen Industriegebiet*, 1919, p. 3.
6 *Ibid.*, pp. 3–4.
7 von Oertzen, *Betriebsräte in der Novemberrevolution*, p. 128.
8 Neunerkommission, *Sozialisierung des Bergbaues*, pp. 8–9.
9 *Ibid.*, p. 9.
10 *Ibid.*, p. 12.
11 *Ibid.*, p. 19. Bold in original.
12 P. von Oertzen, 'Die Grossen Streiks der Ruhrbergarbeiterschaft im Frühjahr 1919. Ein Beitrag zur Diskussion über die revolutinäre Entstehnungsphase der Weimarer Republik', *Vierteljahrshefte für Zeitgeschichte*, 6:3 (1958), 231–6.
13 Neunerkommission, *Sozialisierung des Bergbaues*, p. 5.
14 *Ibid.*, p. 20.
15 *Ibid.*, pp. 24–8.
16 *Ibid.*, p. 25.
17 *Ibid.*, p. 25.

18 *Freie Arbeiter-Union. Organ der Schacht- und Betriebsorganisation der Hand- und Kopfarbeiter Deutschlands/Wirtschaftsbezirk Rheinland-Westfalen*, 30 October 1920. The *Union* claimed that its paper had 126,000 subscribers, *Polizeipräsident* Bochum – *Regierungspräsident* Arnsberg, 8 October 1921, STAM Reg. Arnsberg 14442.
19 *An die revolutionären Arbeiter*, 21 October 1921, STAB, LA 1314.
20 *Polizei Direktor* Bochum to *Regierungspräsident* Arnsberg, 2 July 1921, STAM Reg. Arnsberg, 14442. Individual members were, however, permitted to affiliate within any party.
21 *Polizei Direktor* Bochum to *Regierungspräsident* Arnsberg, 21 June 1921, STAM Reg. Arnsberg 14442. The Bolsheviks were also condemned for their suppression of Russian Syndicalists.
22 F. Oerter, *Was wollen die Syndikalisten?* (Berlin: Verlag 'Der Syndikalist', 1920), p. 11.
23 *Ibid.*, p 16.
24 See for example F. Barwich, *Der kommunistische Aufbau des Syndikalismus im Gegensatz zum Partei-Kommunismus und Staatsozialismus* (Berlin: Der Syndikalist, 1920); Oerter, *Was wollen die Syndikalisten?*; F. Brandt, *Syndikalismus und Kommunismus* (Berlin: August, 1919).
25 Oerter, *Was wollen die Syndikalisten?*, p. 11.
26 See Ch. 2.
27 *WK*, 15 March 1925.
28 *Ibid.*, 26 March 1925.
29 See *An das arbeitende Volk Deutschlands*, 14 March 1925, *Ihr müßt wahlen?*, 1925, *Arbeiter! Parteigenossen! Proletariat von ganz Deutschlands!* 1919, IZF, Mappe 10.
30 *Ebert oder Ludendorff?* 1919, IZF, Mappe 10.
31 *Gegen das Betriebsrätegesetz. Für revolutionäre Betriebsräte*, 1919, IZF, Mappe 10.
32 *An die Arbeiter der ganzen Welt* and *Die russische proletarische Revolution Sowjet-Rußland ist in Gefahr!* 1919, IZF, Mappe 10.
33 O. Hue, *Die Sozialisierung der Kohlenwirtschaft* (Berlin: Vorwärts Verlag, 1921), p. 14.
34 *BZ*, 1 February 1919.
35 *WM*, 14 February 1919; *BZ*, 7 October 1916.
36 *Fachgruppe Bergbau*, press release, 20 November 1920, BBA 230; Minutes of Directors Meetings Powell Duffryn Steam Coal Company Ltd, 18 March 1924, GRO D/D PD 1/8, p. 20; *CG*, 17 January, 23 March and 19 December 1919; Ferndale Lodge Minutes, 27 January 1926, L/33/A7; *AL* 25 October 1924.
37 Minutes of MFGB, Executive Committee Meeting, 13 January; Special Conference, Southport, 14–16 January 1919, p. 9 and p. 47.
38 Cited in Arnot, *South Wales Miners*, vol. 2, p. 153.
39 See James, 'War and industry'.
40 Arnot, *South Wales Miners*, vol. 2, p. 143.
41 *WM*, 10 October 1919.
42 F. Hodges, *Nationalisation of the Mines* (London: Leonard Parsons, 1920), p. ix.
43 *Ibid.*, p. 111.
44 *Ibid.*, p. 131.
45 *Ibid.*, p. 81.
46 *Ibid.*, p. 130.
47 F. Hodges, *My Adventures as a Labour Leader* (London: George Newnes, 1925), p. 23.

48 MFGB Minutes, Special Conference, London, 26 February 1919, pp. 135–47.
49 *Ibid.*, p. 66. For a sympathetic account of Hodges see C. Williams, 'The Odyssey of Frank Hodges', *Transactions of the Honourable Society of Cymmrodorion*, 5 (1999), 110–30.
50 *BZ*, 14 January 1919.
51 *Bericht der Sozialisierungskommission über die Frage der Sozialisierung des Kohlenbergbaues vom* 31 July 1920 (Berlin: Engelmann, 1920), pp. 36–40.
52 *Protokoll der Verband der Bergarbeiter Deutschlands, XXI. Generalversammlung, Bielfeld*, 15–21 June 1919, pp. 389–400.
53 H. Löffler, 'Die Sozialiserung des Bergbaues', Nachlaß Löffler, IfSB, Div. 2 Mappe 10, 1920.
54 Hue, *Sozialisierung der Kohlenwirtschaft*, p. 15.
55 *Bk*, 30 November 1918.
56 See *Bk*, 22 and 29 March 1919, 20 March 1920.
57 *Bericht über das 25 jährige Jubelfest und Potokoll der 15. Generalversammlung des Gewerkvereins christlicher Bergarbeiter Deutschlands*, 24 to 27 August 1919, Essen, p. 141 and p. 144.
58 *Ibid.*, 19 August 1922.
59 *Ibid.*, 15 May 1920.
60 H. Imbusch, 'Sozialisierung des Bergbaues', *DA*, 1919, p. 64.
61 *Ibid.*, p. 65.
62 *Ibid.*, p. 66.
63 *WVZ*, 10 January 1919.
64 *Bk*, 18 February 1922.
65 See Imbusch, 'Sozialisierung des Bergbaues', pp. 304–9; Bericht über das 25 jährige Jubelfest, pp. 170–81; H. Vogelsang, *Reichsarbeitsgemeinschaft für den Bergbau* (Essen: Gewerkverein christlicher Bergarbeiter Deutschlands, 1920).
66 *Bk*, 12 April 1919.
67 *Ibid.*, 27 January 1923.
68 See *Bk*, 1 December 1923, 26 January, 1 March, 15 March, 12 April 1924.
69 *Ibid.*, 17 May and 24 May 1924.
70 *Ibid.*
71 R. Hilferding, *Die Sozialisierung und die Machtverhältnisse der Klassen* (Berlin, 1920).
72 *Ibid.*, p. 10.
73 *Ibid.*, p. 20.
74 *Ibid.*, p. 27.
75 Hue, *Sozialisierung der Kohlenwirtschaft*, p. 11.
76 *BZ*, 13 September 1919.
77 Hue, *Sozialisierung der Kohlenwirtschaft*, p. 8.
78 D. J. Peukert, *The Weimar Republic: The Crisis of Classical Modernity* (London: Penguin, 1991), p. 151.
79 *EAZ*, 18 January 1919, *RWAZ*, 16 January 1919.
80 *Vb*, 2 January 1919.
81 *Ibid.*, 4 January 1919.
82 See Mappe 11, IZF.
83 *Vb*, 15 January 1919.
84 Hilferding, *Sozialisierung und die Machtverhältnisse der Klassen*, pp. 15–16.
85 *Ibid.*, 3 June 1920.

86 *RWAZ*, 4 May 1922.
87 Evans, *The German Center Party*, pp. ix–xi.
88 *Ibid.*, pp. 262–3.
89 *EVZ*, 7 January 1919.
90 *Ibid.*, 4 January 1919.
91 *WVZ*, 31 January 1921.
92 *T*, 1 May 1924.
93 *Vb*, 4 January 1919.
94 *Ibid.*, 11 January 1919.
95 *EAZ*, 14 March 1925.
96 *Ibid.*, 21 March 1925.
97 *Vb*, 14 February 1919.
98 *Ibid.*, 16 February 1919.
99 *Vb* 29 April 1924.
100 *Ibid.*
101 Alter Verband, *Unternehmer und Kommunisten während der Bergarbeiterkämpfe im Mai 1924* (Bochum: August, 1924), p. 10.
102 *Ibid.*, pp. 27–8. For details of the dispute see Hans Spethmann, *Der Maistreik 1924 im Ruhrbergbau. Ein grundsätzlicher Arbeitskampf* (Berlin: Reimar Hobbing, 1932); Tschirbs, *Tarifpolitik im Ruhrbergbau*, pp. 262–6; Liu, *The Chameleon State*, pp. 128–9; and H. Mommsen, 'Soziale und politische Konflikte an der Ruhr 1905 bis 1924', in H. Mommsen (ed.), *Arbeiterbewegung und industrieller Wandel. Studien zu gewerkschaftlichen Organisationsproblemen im Riech und an der Ruhr* (Wuppertal: Peter Hammer, 1980), p. 82.
103 Alter Verband, *Unternehmer und Kommunisten während der Bergarbeiterkämpfe im Mai 1924*, pp. 35–6.
104 *Ibid.*, p. 56.
105 *Ibid.*, pp. 56–7.
106 *Ibid.*, pp. 61–2.
107 *Ibid.*, p. 76.
108 *WM*, 24 April 1919.
109 The pamphlet was designed as a sequel to *The Miners' Next Step.*
110 W. F. Hay, *Industrial Democracy for Miners: A Plan for the Democratic Control of the Mining Industry* (Porth: D. A. Davies, 1919), p. 2.
111 *Ibid.*, p. 4.
112 *Ibid.*, p. 7.
113 *Ibid.*, p. 5.
114 *Ibid.*, p. 8.
115 For details of the dispute see, Arnot, *South Wales Miners*, vol. 2, pp. 189–95; Cole, *Labour*, pp. 157–61.
116 MFGB Minutes, Special Conference, London, 30 September 1920, pp. 1249–68.
117 *LV*, 15 May 1920.
118 *WM*, 15 November 1920. Stead, 'Vernon Hartshorn', p. 93.
119 *LV*, 13 April 1918.
120 *WM*, 5 and 6 July 1921, and 9 July 1921.
121 See Ch. 4.
122 J. Thomas, *The Miners' Conflict with the Mineowners*, Personal Papers: Dan Evans SWCC: MNA/PP/21/3, pp. 68–9.
123 *CWM*, February 1923, p. 27.

124 *Ibid.*, March 1924, p. 63.
125 See Arnot, *South Wales Miners*, vol. 2, pp. 224–6; Francis and Smith, *The Fed*, pp. 30–1; Davies, *A. J. Cook*, pp. 60–3; Griffiths, *Davies*, pp. 65–7. Davies attended the Second World Conference of the RILU.
126 *CWM*, September 1923, pp. 213–17.
127 *Union. Zentral-Organ der Union der Hand- und Kopfarbeiter*, 15 September 1921. See also von Oertzen, *Betriebsräte in der Novemberrevolution*, p. 214.
128 *BZ*, 2 December 1922.
129 For the latter see Ness Edwards's contribution, 'Call to the Miners', *All Power*, February 1923.
130 *LV*, 29 May 1920.
131 F. Hodges, *Workers' Control in the Coal Mining Industry* (London: The Mines for the Nation Campaign Committee, 1920), p. 2.
132 *Ibid.*, p. 3.
133 *WM*, 19 August 1919.
134 For examples see pamphlet no. 1, 'How The Miner Pays!' and no. 7, 'The Coal Industry: Its Rise ad Decline', 1919, TUC Library, HD 9551.6.
135 'The Mines for the Nation', pamphlet no. 10 'The Coal Scuttle', and no. 9 'Coal Nationalisation and the Middle Class', TUC Library HD 9551.6.
136 See *Further Facts from the Coal Commission being a history of the second stage of the coal industry commission with excerpts from the evidence*, 1919, pp. 17–18, TUC Library, HD 9551.5.
137 'The Mines for the Nation', pamphlet no. 2, 'Wages ... Prices ... Profits', 1919, TUC Library, HD 9551.6.
138 'The Mines for the Nation', pamphlet no. 4, 'How Nationalisation will promote Enterprise', 1919, TUC Library, HD 9551.6.
139 'The Mines for the Nation', pamphlet no. 1, 'How the Miner Pays!', TUC Library, HD 9551.6.
140 *LV*, 14 May 1921.
141 *The Times*, 6 September 1926 and 16 October 1926, clippings, PRO, Lab 27/5.
142 *MP*, 7 May 1921.
143 *CG*, 27 June 1919.
144 McKibbin, 'Classes and cultures', p. 157.
145 General Minute Book of MSWCOA, 22 March 1919, NLW, MG 14, p. 87.
146 Minutes of the Powell Duffryn Steam Coal Company Ltd., AGM, 20 March 1920 and 18 March 1924, GRO D/D PD 1/7 and D/D PD 1/8, p. 140 and p. 30 respectively.
147 *Ibid.*, AGM, 14 March 1922, GRO D/D PD 1/7, p. 254.
148 *WM*, 12 May 1919 (Consolidate Cambrian Ltd), 2 July 1919 (Newport-Abercarn Black Vein Steam Co Ltd), 3 July 1919 (Powell Duffryn Steam Coal Company Ltd.), 8 July 1919 (Tredegar Iron & Coal Co.).
149 *WM*, 28 January 1919.
150 See S. Berger, 'Working-class Culture', p. 30 and M. Lieven, 'A "new history" of the south Wales coalfields', *Llafur*, 8:3 (2002), 94.
151 *All Power*, August 1923.
152 *LV*, 1 January 1921.
153 B. Ziemann, 'Das Fronterlebnis des Ersten Weltkrieges – eine sozialhistorische Zäsur? Deutungen und Wirkungen in Deutschland und Frankreich', in H. Mommsen (ed.), *Der Erste Weltkrieg und die europäische Nachkriegsordnung.*

*Sozialer Wandel und Formveränderung der Politik* (Cologne: Böhlau, 2000), pp. 79–80.

154 See for example the legend of the Essen water tower: E. Lucas-Busemann, 'Kapp Putsch und Rote Ruhrarmee', in Gorlas and Peuket (eds), *Ruhrkampf 1920*, pp. 69–70.

155 M. Gailus, '"Seid bereit zum roten Oktober in Deutschland." Die Kommunisten', in D. Lehnert and K. Megerle (eds), *Politische Identität und nationale Gedenktage. Zur politischen Kultur in der Weimarer Republik* (Opladen: Westdeutscher Verlag, 1989), p. 88. Weitz, *Communism*, pp. 178–85.

156 *Oberbürgermeister* Hamborn to *Regierungspräsident* Düsseldorf, 1 June 1921, STAD 26120.

157 *An die Bergarbeiter des Ruhrgebietes*, 1925, STAM, 14455.

158 *BZ*, 28 February and 7 March 1925.

159 Weitz, *Communism*, pp. 263–70.

160 *Ibid.*, pp. 196–202.

161 Horner Personal Papers, SWCC, MNA/PP/46/35.

162 For the example of Gwaun-cae-Gurwen see, *LV*, 15 September 1921 and 28 January 1922.

163 McKibbin, 'Classes and cultures', pp. 157–8.

164 Williams, *When was Wales?*, p. 238.

165 *WO*, April 1918, pp. 113–4.

166 *PO*, 11 November 1922. Ironically Williams was a Liberal candidate.

167 *AVCECN*, 5 December 1918.

168 *Ibid.*, 12 December 1918.

169 *FPM*, 3 November 1922.

170 *Ibid.*, 24 October 1924.

171 K. O. Morgan, *Rebirth of a Nation: A History of Modern Wales* (Oxford: Oxford University Press, 1981), p. 193.

171 *Cymmer Searchlight*, 23 October 1925, SWCC, SC 461.

173 *CWM*, November 1924, p. 275.

174 *SWA*, 21 October 1924.

175 *AVCECN*, 29 November 1923.

176 *GFPRL*, 30 November 1923.

177 *GCT*, 18 October 1924.

178 *AVCECN*, 9 November 1922.

179 *CJ*, 30 November 1923.

180 *GFPRL*, 30 November 1923.

181 *PO*, 24 November 1923.

182 *GCT*, 7 December 1918.

183 *Ibid.*, 14 December 1918.

184 *AL*, 4 November 1922.

185 May, 'Charles Stanton', pp. 145–66.

186 M. Lieven, 'A fractured working-class consciousness?', pp. 729–56.

187 See A. R. Griffin, *The Miners of Nottinghamshire 1914–1944: A History of the Nottinghamshire Miners' Unions* (London: George Allen & Unwin, 1962).

188 See Francis and Smith, *The Fed*, pp. 91–2; Horner, *Rebel*, pp. 130–1.

# Conclusion

This study of trade unionism and political development ends with the defeat of the MFGB in 1926. This is, admittedly, a somewhat arbitrary point at which to conclude. The trade unions and political parties examined here continued to reproduce the identities they had created through newspapers, magazines and speeches beyond this point. Perhaps a more obvious finishing point would be 1933 and the Nazi *Gleichschaltung* (co-ordination) which set about destroying civil society and suppressing other identities. This, however, would ignore the SWMF and Labour Party's continued efforts to forge an identity in the 1930s. In fact any endpoint seems unsatisfactory. Rather than being marked by sharp breaks, the construction of identity in both regions was an ongoing process.

This study concludes in 1926 because the hypothesis that underpinned it no longer seems wholly applicable after this point. From 1890 up until the mid-1920s the development of the miners' movements in the two regions had demonstrated divergent tendencies. In the Ruhr the miners' movement was generally characterised by fragmentation; in south Wales by homogenisation. The *Verband zur Wahrung und Förderung der bergmännischen Interessen im Rheinland und Westfalen* initially united the miners of the Ruhr, but over the next two decades the union movement fractured on lines of religion and ethnicity. By 1914 the *Alter Verband* competed with the *Gewerkverein* and Polish ZZP. The Great War encouraged greater cooperation between the two largest unions, but the revolutionary period saw further fragmentation as various syndicalist and communist unions emerged. In south Wales, an initially fragmented trade unionism was united following the 1898 strike. Unlike the *Gewerkverein* and the *Alter Verband*, the SWMF was also able to bolster its position during the First World War and become almost an industrial union.

Similarly, in political terms this characterisation of homogenisation in south Wales and fragmentation in the Ruhr is generally applicable. In the 1890s the two-party system prevalent in the Ruhr became a three-party one as the SPD emerged from the restrictions imposed on it by the Anti-Socialist Law. During the war the socialist-democratic movement split, laying the foundations for competition between the Communists and Social Democrats under Weimar.

Before the war the Liberals largely dominated the south Wales mining valleys. However, cracks in their control were becoming evident at local level in the 1900s, while the concern with 'labour' and 'class' as independent political entities was also growing at parliamentary level. The war boosted the political labour movement both organisationally and psychologically. It revitalised institutions such as the TLCs, while demonstrating that those from working-class origins were competent administrators. This provided the basis for the Labour breakthrough in the 1920s, when it superseded the Liberals as the party of the mining valleys.

After 1926, however, the process of homogenisation in industrial terms broke down. The SWMIU was formed and until late in the 1930s the SWMF struggled to remove its rival. On the other hand, Labour consolidated its hold on the mining valleys. Thus homogenisation on both political and industrial levels no longer paralleled each other. In the Ruhr the picture typically was more complicated. The disbanding of the *Union* in the Ruhr appeared to mark an end to the progressive fragmentation in miners' trade unionism there. But in reality its disappearance coincided with the Communists' decision to infiltrate social-democratic bodies, a practice that had been pursued in an unsystematic way before 1924. The picture of a relatively more united movement was therefore deceptive. The adoption of the 'third phase' policy towards the Social Democrats exposed the enduring nature of the divisions in the socialist labour movement. This division found organisational expression in the appearance of the *Revolutionäre Gewerkschafts-Opposition* (RGO, Revolutionary Trade Union Opposition). By 1931 the RGO succeeded in winning 29 per cent of the vote in the works' council elections.[1] Meanwhile, in the political sphere, fragmentation continued, especially as the National Socialists began to make their first inroads into the region.

However, it is not the purpose of this book to examine why these processes broke down by 1926. Instead it has sought to answer why they existed at all and why they differed. As noted in the introduction, both south Wales and the Ruhr were in many respects atypical of other coalmining regions within Britain and Germany. Certainly they shared some characteristics with other industrial regions and through national organisations were aware of supraregional and international events, trends and currents of thought. Yet in both countries each coalfield had its own patterns of industrial development, ownership, working practices, immigration, settlement and mechanisation as well as its own political traditions. These factors influenced both trade unionism and party politics, but not necessarily directly. Often their influence was mediated by the active attempts of politicians and union leaders to provide an interpretative framework to explain them as encapsulated in their discursively constructed identities. As such, south Wales and the Ruhr cannot be considered microcosms of the British and German working class respectively, since in other areas working-class leaders would have had to forge identities that sought to explain different social and economic developments. Rather they represented one part of the mosaic that comprised national working-class movements.

Bearing this caveat in mind, this work has tried to provide a theoretical framework that can be applied to other industrial areas. It has tried to account for these differences between the south Wales and Ruhr coalfields by using a three-part framework.

First, it has argued that the lifeworld of the Welsh miners was more synchronous than that of the Ruhr miners. In doing so it has sought to integrate some of the theories used to explain solidarity and strike propensity among miners with the concept of milieux widely applied in German historiography. The comparison of south Wales and the Ruhr reveals that the former conforms more closely to Bulmer's 'ideal-type traditional mining community'. This difference is apparent on several levels. First, although one must be cautious not to overestimate the physical isolation of the mining communities, it is clear that the Ruhr, as the premier industrial district in Germany, differed from south Wales, which was one of several important coalfields. The Ruhr also contained several important cities, while in south Wales the largest conurbations lay outside the coalfield at the point of export. Secondly, mining was economically predominant in south Wales, while the industrial base in the Ruhr was more varied. In the latter region, steel, iron and armaments factories represented alternative sources of employment. Thirdly, mining in both areas was dangerous, but the memory of their previous privileged position under state control reinforced perceptions among the Ruhr miners that mining as an occupation had been downgraded since the liberalisation of the industry. In south Wales, on the other hand, it appears the occupation was regarded more positively as many miners refused promotion to deputy level even when they had gained the necessary qualifications. Fourth, the dominance of mining made the occupational structure of south Wales more uniform than in the Ruhr. It also appears that the Ruhr miners had somewhat less autonomy at work and were greatly affected by mechanisation and rationalisation in the 1920s. By contrast, in south Wales there was a greater continuity in the experience of work pre- and post-war. Fifth, the communal activities of the miners differed. The Welsh miners did engage in activities that were community focused such as rugby and choral singing. The Ruhr miners also had a rich club life, but rather than represent the community, it was often intimately linked to a certain milieux or subculture. Sixth, both coalfields shared a sharp gender segregation and division of work, but the 'hereditary' nature of mining as an occupation was greater in south Wales. Bulmer's seventh point, an economic/political conflict of interest between owners and miners, applies to both regions, but it is evident that the SWMF was more successful in achieving its aims than either the *Alter Verband* or *Gewerkverein* were. Taking these points together it is clear that south Wales conforms more closely to the 'ideal-type'.

The idea of milieu further illustrates the fragmented nature of the Ruhr miners' lifeworld. German historiography has made much use of this concept to differentiate between sections of the mining workforce and it would seem possible to integrate this with Bulmer's model, especially with respect to points five and seven. Social-democratic, Catholic and Polish milieux have been described.

Each was made up of a network of associations that variously included churches, political parties, trade unions, educational societies, sporting teams, co-operatives and drinking clubs. Apolitical or neutral associations (i.e. not integral to a particular subculture) did exist, but during this period these milieux fractured the lifeworld of the Ruhr miners. Of course, various competing or antagonistic associations existed in south Wales, but they did not equal the almost all-embracing nature of the milieux in the Ruhr. Indeed, the SPD set out to create a subculture that could serve its members' needs from 'cradle to grave'.

In fact, the concept of milieux as applied to German history seems inappropriate for south Wales during this period. The theoretical literature is clear on the pivotal role state pressure played in consolidating milieux. The *Kulturkampf*, the Anti-Socialist Law and the 'Germanisation' policy represent prime examples of this process as they served to make certain groups (Catholic, Social Democrat, Pole) more conscious of their interests and encouraged them to establish bodies dedicated to protecting them. Any clear distinction between the milieux and the parties would be a false dichotomy. The parties were integral parts of the milieu; they were its political expression. The Centre Party, for example, emerged to counter the effects of the *Kulturkampf*. As already noted, the Centre represented the 'classic' milieu party. The SPD was less successful in its attempts to create its own milieu, but it did establish a sub-culture made up of former Protestant cottage miners and some sections of the immigrants.[2] Despite this the SPD was able to challenge the Centre and National Liberals to form a three-party political system in the 1890s. In the 1920s, the political landscape became even more fragmented by the emergence of the KPD. Like the SPD in the 1890s, the KPD also sought to build its own network of ancillary organisations (or to usurp them from the SPD).

Such 'top-down' pressures were not evident in south Wales during the same period. The mining valleys were not free from ethnic tensions, as illustrated by anti-Irish and anti-Jewish riots. There were also physical confrontations between English and Welsh. Ethnic and religious associations were also established. Significantly, these rivalries did not lead to organisational enmity. Admittedly, the Irish formed their own societies but by the late 1890s the Irish community was more integrated into Welsh communities. In any case, the mining workforce in south Wales was less ethnically diverse than in the Ruhr. Miners were generally Welsh or English. English immigrants were easier to assimilate than non-German-speaking Poles. The large number of English immigrants also adapted the existing communities to their needs. While Poles in the Ruhr found it difficult to find Polish-speaking priests to administer services, many chapels switched from using Welsh to English. The acceptance of English as the *lingua franca* in much of the Welsh coalfield is illustrative of the comparative ease with which migrants were accepted into (and changed) the mining communities. That it occurred without laws purposively designed to 'Anglicise' the Welsh was probably its strength. Official attempts to 'Germanise' the Poles backfired and merely reinforced their milieu. Furthermore, unlike the Ruhr, where the population was roughly equally divided between Protestantism and Catholicism, in south Wales the people were overwhelmingly Nonconformist. Thus divisions on religious grounds also lacked the force they had in the Ruhr.

The issue of milieux (or the lack of them) relates to the second explanation used to explain the differences between the two coalfields, that of civil society. It is clear that civil society in the Ruhr conforms to the rigid, segmented societies described by Rosenblum and Post. There are numerous examples of how dual membership of organisations was not tolerated. For example, membership of the *Gewerkverein* was dependent on an acceptance of Christianity and the open rejection of social democracy. Indeed, many in the Catholic Church saw the Christian trade union movement as an antidote to social democracy. Naturally, Polish organisations expected their members to be Poles. The *Alter Verband* claimed to be neutral on political and religious issues, but it is questionable how seriously the miners would have taken such protestations since one of the trade union's leading lights, Hue, stood as an SPD deputy for Bochum. Similarly, Heinrich Imbusch stood for the Centre after the war. The organs of the two largest trade unions, *Der Bergknappe* and *Die Bergarbeiterzeitung*, printed articles that clearly favoured the political positions of Christian-socialism and social democracy respectively. Finally, after the war and the Revolution, dual membership of the SPD and the KPD was incompatible. Both parties were locked in a struggle that fatally undermined the labour movement in the Weimar Republic.

By contrast, in south Wales, dual group membership was generally accepted and tolerated. Individual members of the SWMF could, for example, support whichever political party they desired, despite the affiliation to Labour. Although many ministers were hostile to socialism, opposition from the religious establishment was not monolithic. A significant (and vocal) minority of ministers was supportive of Labour or even socialism. It is significant that Nonconformist journals and the Labour press itself debated the pros and cons of socialism. Even employers were allowed to put their viewpoint forward, as did D. A. Thomas in the *South Wales Worker*. Finally, after the war, plural membership of the CPGB and Labour was also accepted during this period. Communists like Arthur Horner did not regard the CPGB as a rival to Labour, but as a watchdog, dedicated to ensuring that the party did not become complacent or co-opted into the establishment. In the *Colliery Workers' Magazine* articles praising the Russian experiment stood alongside appeals to strengthen the organisation of Labour.

This summary of civil society in south Wales and the Ruhr naturally leads to the third and final part of the framework: discursively constructed identities. The foregoing account may seem an adequate explanation for the differences in trade union and political development. The various trade unions and parties could be explained away as expressions of the various milieux and sub-cultures, or in south Wales the lack of a milieu. Undoubtedly, this represents part of the answer. The lifeworld and civil societies provided the background against which the trade unions and political parties examined here worked. But if these organisations were merely the expression of different social structures, how are the incidences of cooperation in the Ruhr to be explained? Strikes like those that occurred in 1889 and 1905 demonstrated the capacity for the Ruhr miners to show solidarity. Moreover, in south Wales the lack of milieux did not automatically mean a united labour movement. Indeed, the early 1890s were characterised by local unions and conflict between the supporters of the Sliding Scale and the MFGB. Even after the formation of the SWMF tensions centred on older leaders, who advocated collaboration with employers, and a younger generation of leaders, or 'advanced' men, who believed that the union and the owners were inevitably locked in conflict. Why were the *Alter Verband* and the *Gewerkverein* unable to sustain cooperation, while the SWMF was able to contain potential divisions?

The SWMF sought to base its identity on the miners as a whole. Its identity was 'community-fixated' in that it focused on the miner as a working man rather than on a particular ideal. Therefore, its activists, whatever their differences, sought continually to emphasise solidarity within their discourse, even when there was bitter division among leaders over policy. It was policy (i.e. what was best for the union and the miners) that was at the crux of conflict, rather than what the miners should be. During times of industrial conflict this identity would be placed in juxtaposition to the employers, not to other miners, unless blacklegs. During times of industrial peace the union focused on contrasting the dignity and honesty of those who were members with the 'feckless' nature of those who were not. Since this discursive practice was essentially focused on a miner/employers or organised/unorganised dichotomy it could incorporate a wide range of elements, provided they maintained the central paradigm. This, coupled with a more fluid civil society that accepted plurality, meant a more inclusive identity could be constructed. The identity of the miner constructed by the SWMF was, therefore, comparatively unproblematic.

Conversely, the *Alter Verband*, *Gewerkverein* and ZZP based their identities on specific conceptions of the miner. The 'milieux-fixated' identities created by the unions, based on Christian-socialism, Marxism and Polish ethnicity respectively, were comparatively doctrinaire. By joining these unions a miner was declaring he was a certain kind of man. Even the names of the *Gewerkverein* and ZZP demonstrated that they sought not to represent the miner, but a type of miner. The unions also based their organisational discourse on demonstrating what they were not. As already noted, the *Gewerkverein* explicitly saw part of its function as the combating of social democracy. This separation and juxtaposition of the identities of Christian and social-democratic miners was crucial. To abandon it would seem to represent an end to the union. This fear determined the *Gewerkverein* reaction after any period of cooperation and it vigorously rejected any suggestion of amalgamation. Thus discursive identities could condition the trade unions' actions. The discursive practices of the trade unions therefore not only drew upon pre-existing social divisions, but also served to reinforce them. Thus, although the unions sought to broaden their appeal to embrace as much of the workforce as possible, their ability to recruit was intrinsically limited by their inflexible identities.

The political parties differed slightly from the unions. Their identities were even more explicitly drawn as each sought to distinguish itself from its rivals. The Liberals based their appeal on an emotive mixture of nationalism, Nonconformism and community identity. The Liberals combined these elements to form a 'community-fixated' discourse that contrasted their idealised notion of the *gwerin* with the Anglicised, Anglican landowners. Within Liberal political language the landlords were excluded from the community. On the other hand, by voting Liberal the miners could demonstrate that they belonged to a particular Liberal vision of the community.

The concept of 'class' was not often used by the Liberals in the early 1890s. When it was employed it was used pejoratively and meant to imply sectionalism and vested interests and contrasted to the more encompassing idea of the 'people'. The Lib-Labs, however, introduced a positive idea of class, one that referred to the workers. As the example of Alfred Thomas demonstrates, Liberal discourse was also able, up to a point, to accommodate growing concerns with the representation of 'labour' and 'class' in the late nineteenth and early twentieth century.

Unfortunately for the Liberal Party, the discourse of class became increasingly focused on its independent representation. Liberal discourse emphasised cooperation between labour and capital and its vision of the community encompassed both the miners and the coalowners. Within Labour discourse, on the other hand, concepts of community and the working class were increasingly elided. This was especially evident in Keir Hardie's rhetoric. He used many elements of Liberal political language to great effect, but for him it was mainly the workers who made up the community. Capital was excluded and portrayed as a threat. This trend was reinforced by the First World War, which also demonstrated that Labour was a competent administrator. The industrial disputes of the 1920s strengthened this tendency and the issue of nationalisation allowed a particularly apt expression of this labour/capital dichotomy. Yet, part of Labour's success in the inter-war period was down to its adoption of elements of Liberal discourse. In the elections immediately after the war older radical issues such as Disestablishment remained prominent. By accepting these ideas into its own political language Labour was able not merely to emerge as a potential challenger to Liberalism, but to adopt the mantle of 'progressivism' and eventually dominate the mining valleys.

In the Ruhr, the National Liberals and Centre parties, like the Liberal Party in Wales, also combined nationalism and religion within their discourse. They sought to portray themselves as *Volkspartei*, representing the community as a whole. Both adopted a 'friend-of-the-workers' approach. However, like the trade unions in the Ruhr, their discourse was 'milieu-fixated'. The extent of their appeal was limited by their concentration on a specific religious group. Furthermore, like the British Liberals, both the Centre and the National Liberals were reluctant to adopt candidates from working-class backgrounds. The struggle between the worker Gerhard Stötzel and a middle-class member of the Centre Party in 1877 mirrored the electoral contest that Mabon faced against a

Liberal mineowner in 1885. But class for the Centre was bound up with issues of confession. Since the party had been formed to protect Catholic interests it could not move away from its essentially Catholic image either during the *Kaiserreich* or during the early years of the Republic. Just like the *Gewerkverein*, the Centre Party in the Ruhr predicated its discourse on the idea of the Christian, specifically Catholic, miner. This necessarily undermined its attempt to appear as a broad people's party.

The SPD self-consciously adopted a different approach. It did not predicate its identity on a specific religious community. Indeed, it was castigated by both the Centre and the National Liberals as being both irreligious and a party of *Reichsfeinde* (enemies of the state). Unlike their opponents, the SPD adopted an identity that was explicitly predicated on the working class. On this basis they accused the Centre and National Liberals of being employers' parties. This identity was partially successful in drawing support from Protestant *Kötterbergleute* (cottage miners), disaffected with the effects of economic liberalisation, and some of the immigrants to the region. But it was undermined by its very attitude towards religion. Hue, for example, constantly maintained that religion was a private matter, not a political one. But for the Centre and many Catholics the issue was anything but private. It was a public matter and issues such as denominational schools repeatedly surface. The irreligious element in the SPD's political language hampered its appeal, just as Catholicism limited that of the Centre. Also its heavy emphasis on class meant the SPD was largely unsuccessful in reaching out to other groups in Wilhelmine society. In the 1900s the party struck a more patriotic note, but its primary identity remained based on class. This was evident in the way in which it constantly rejected the idea that the Centre could do anything of note for the workers. In its own language the SPD was the only workers' party. Using this identity the SPD was able to win several electoral contests in the Dortmund and Bochum seats, but was unable to dominate the regions due to the entrenched nature of Centre and National Liberal support.

Ultimately, the SPD made its breakthrough in 1919 under a very different electoral system. Unfortunately, for the SPD this triumph proved short-lived. While the Great War had strengthened the Labour Party, it had accentuated divisions within the SPD, which split into the USPD and MSPD. In 1919 the USPD attracted little electoral support, but in 1920 their share of the vote increased at the expense of the SPD. Later, following the re-unification of the MSPD and part of the USPD, the KPD inherited much of this support. Moreover, the KPD usurped much of the class discourse that had been the sole preserve of the SPD before 1914. In effect it was drawing on older, more radical notions of class that the SPD had abandoned for a more reformist path in the 1900s. The KPD discourse was also 'milieu-fixated' in that it drew the majority of its membership from the unemployed. Excluded from the workplace, the KPD increasingly focused on violent confrontation and the struggle to protect Soviet Russia.

The SPD meanwhile found itself in the unfamiliar position of facing a serious challenger on the left. It also had to contend, on the one hand, with the dictates of power and, on the other, with the expectations of its followers. The disillusionment that many felt in the 1920s benefited the KPD, who capitalised on notions of betrayal of the Revolution. Thus, the idea of the *Arbeiterpartei* that had been the SPD's mainstay before the war became contested. Yet the SPD could not abandon class as a defining feature of its discursive identity. Its very idea of the Republic was a socialist one. Its attempt to build upon the patriotic element evident before the war was not wholly successful for the same reason that the Centre was unable to convincingly claim that it was a *Volkspartei.* Class was too ingrained in its identity to be easily ejected. To do so risked losing more of the party's core constituency to the Communists.

This book has argued that an understanding of the discursive practices of the trade unions and political parties is crucial for explaining differences in party political and trade union development in the coalfields. Within a context of a comparatively more synchronous lifeworld and a fluid civil society, the SWMF, Liberal party and later Labour developed in south Wales more inclusive identities. It was the inclusiveness of these identities that allowed these organisations to achieve the dominant positions they did. Indeed, Liberalism began to falter when it reached the limit of its ability to encompass different discourses. Its ability to accommodate 'labour' and 'class' did not extend to independent representation and the increasing emphasis laid on these terms undermined the Liberals' position. This trend was paralleled by changes to what the community was perceived to include. Coalowners, who had been seen by Liberalism as an integral part of the community, were firmly excluded in Labour's political language. Thus although Labour focused on class, class by now meant much the same as the community.

In the Ruhr, the 'milieu-fixated' nature of discursive identities meant that they were exclusive. A fragmented lifeworld and segmented civil society made it difficult for a community discourse to be formed. Instead each identity, forged through discourse, claimed to represent a certain group. The trade unions and the parties placed each other's identities in juxtaposition. This contrasting of one identity to another became necessary for the very survival of the organisations involved. Although modifications were made, large changes were difficult to make as they risked alienating the core constituency. These discursive practices limited cooperation and negated amalgamation. It was only the trauma of National Socialism and war that eventually broke these barriers down for good.

## Notes

1 Jäger, *Bergarbeitermilieus und Parteien*, p. 49.

2 Rohe, 'Die Ruhrgebietssozialdemokratie im Wilhelminischen Kaiserreich', pp. 333–4.

# Bibliography


## Primary Sources

### A. British archives

**1. South Wales Coalfield Collection, Swansea University**

*Personal Papers:*

Dan Davies (SWCC MNA/PP/21/2–5)
Mrs Beatrice Davies (SWCC MNA/12/9 and 21)
W. G. Derrick (SWCC MNA/PP/18/4/2–10)
Arthur Horner (SWCC MNA/PP/46/35)
Len Williams (SWCC MNA/PP/129/9)

*Lodge Minutes:*

Rhondda No. 1 District Minutes
Abergorky Lodge Minutes
Blaenavon Lodge Minutes
Coegnant Lodge Minutes
Caerau Lodge Minutes
Cambrian Lodge Minutes
Ferndale Lodge Minutes
Llanhilleth Lodge Minutes
Mardy Lodge Minute
Oakdale Lodge Minutes
Tylorstown Lodge Minutes
Miners' Eight Hours Bill. South Wales Miners' Federation Open Session, 17 March 1908 (SWCC MNA/NUM/3/10/19)

**2. Glamorgan Records Office**

Aberdare UDC Minutes 1895–1926
Maesteg UDC Minutes 1926
Ogmore and Garw UDC Minutes 1924–1926
Reports of H. M. Inspector of Mines for the South Wales/Southwestern District, 1887–1926 (GRO D/D NCB 67/2/1–22)
Minute Book Crynant Collieries Co. Ltd. (GRO D/D PD 27/2)
Minute Book Cwmaman Co. Ltd. (GRO D/D PD 33–4)

Reports of the manager of the Locket Merthyr Collieries (GRO D/D NCB 16/35)
Minutes of directors meetings Powell Duffryn (GRO D/D PD 1/1–8)
Minute Book of the Managing Committee of Powell Duffryn (GRO D/D PD 2/3–5)
Diaries of Mr T. H. Bailey, General Manager of the Plymouth Colliery Co. (GRO D/D NCB 16/37)
Brecon Road Liberal Association, Correspondence (GRO D/D Xes 1–20)
Handbills and political pamphlets, A. C. Fox Davies, 1910 (GRO D/D Vau 23/1–10)
Handbills and political pamphlets, Keir Hardie, 1910 (GRO D/D Vau 24/1–16)
Handbills and political pamphlets, W. Pritchard Morgan, 1910 (GRO D/D Vau 25/1–11)
Keir Hardie election address, 3 December 1910 (SWCC SC 491)
Abercynon Workmens' Institute Records (GRO D/D X 910/2–16)
Diary of Chief Constable Lionel Lindsay, 1898 & 1911 (GRO D/D Con 77/78)
Elwood, Peter, *A History of Temple*, 1987 (GRO D/D Bap 44/1)
Minute Book of Llanbradach Workmens' Hall, 1920–25 (GRO D/D X 816)
Minute Book of the Society for Establishing and Supporting English Congregational Churches In South Wales 1877–1893 (GRO D/D E Cong 1/1)
Rhys, Trefor J., *A History of Capel Rhondda*, 1953 (GRO D/D W Bap 12/1)

**3. Carmarthen County Records Office**
Ammanford UDC Minutes 1903–1927
Cwmamman UDC Minutes 1919–1927

**4. South Wales Miners' Library**
South Wales Miners' Federation Minutes

*Transcripts of Interviews undertaken by the South Wales Coalfield History Project:*
Will Coldrick (SWML AUD 339)
Idris Cox (SWML AUD 171)
Rees Davies (SWML AUD 207)
Trevor Davies (SWML AUD 165)
Edgar Evans (SWML AUD 213)
Glyn Evans (SWML AUD 346)
Mrs J. Evans (SWML AUD 172)
Jim Evans (SWML AUD 221)
Max Goldberg (SWML AUD 347)
James Griffiths (SWML AUD 90)
Harry Howells (SWML AUD 302)
Henry John (SWML AUD 220)
Morgan Jones (SWML AUD 227)
Henry Lewis (SWML AUD 201)
Archie Lush (SWML AUD 276/338)
Tommy Nicholas (SWML AUD 97)
Cyril Parry (SWML AUD 333)
Will Picton (SWML AUD 193)
George Prothero (SWML AUD 309)
Claude Stanfield (SWML AUD 205)
Norman Thomas (SWML AUD 235)
Frank Williams (SWML AUD 169)
John Williams (SWML AUD 228)

J. L. Williams (SWML AUD 396)
William Henry Williams (SWML AUD 290)

**5. National Library of Wales**

General Minute Book of the Monmouthshire and South Wales Coalowners' Association, 1890–1926 (NLW MG5–20)

Minutes of proceedings at meetings between Representatives of Monmouthshire and South Wales Coal Owners' Association and Representatives of the Monmouthshire and South Wales Colliery Enginemen, Stokers and Craftsmen's Association, 1919–20, 1924–16 (NLW MV274–318)

Minutes of proceedings at meetings between Representatives of Monmouthshire and South Wales Coal Owners' Association and Representatives of the South Wales and Monmouthshire Mechanical and Surface Workers' Union, 1921–24 (NLW MV488–97)

**6. Trade Union Congress Library**

'Mines for the Nation' Campaign Leaflets (TUC HD9551.6)

**7. Public Records Office**

Cabinet Papers (PRO Cab/24)
Ministry of Labour Files (PRO Lab 27/1–5)
Industrial Unrest, Commission of Inquiry (Reports) No.7 Division, 1917–18

## B. German archives

**1. Institut für Soziale Bewegungen**

*Personal Papers:*
Nachlaß Heinrich Imbusch (IfSB)
Nachlaß Heinrich Löffler (IfSB)
Nachlaß Weeke (IfSB)

*Trade Union Records:*
*Alter Verband*

Protokoll des Bergarbeiter-Delegiertentages zu Dorstfeld, 18 August 1889, Dorstfeld

Protokoll der 1. Generalversammlung, Verband zur Wahrung und Förderung der bergmännischen Interessen in Rheinland und Westfalen, 20 October 1889, Bochum

Protokoll der 2. Generalversammlung, Verband zur Wahrung und Förderung der bergmännischen Interessen in Rheinland und Westfalen, 27 October 1889, Bochum

Protokoll der 3. Generalversammlung, Verband zur Wahrung und Förderung der bergmännischen Interessen in Rheinland und Westfalen, 8 March 1890, Bochum

Protokoll der 4. Generalversammlung, Verband zur Wahrung und Förderung der bergmännischen Interessen in Rheinland und Westfalen, 11 May 1890, Bochum

Protokoll der 1. Generalversammlung deutscher Bergarbeiter Verband, 1 November 1890, Bochum

Protokoll der 2. Generalversammlung deutscher Bergarbeiter Verband, 19 July 1891, Bochum

Protokoll der 3. Generalversammlung deutscher Bergarbeiter Verband, 31 July 1892, Bochum

Protokoll der 4. Generalversammlung deutscher Bergarbeiter Verband, 20 August 1893, Dortmund

Protokoll der 5. Generalversammlung deutscher Bergarbeiter Verband, 26 August 1894, Bochum
Protokoll der 6. Generalversammlung deutscher Bergarbeiter Verband, 25 August 1895, Bochum
Protokoll der 7. Generalversammlung deutscher Bergarbeiter Verband, 2 August 1896, Bochum
Protokoll der 8. Generalversammlung deutscher Bergarbeiter Verband, 18 April 1897, Helmstadt
Protokoll der 9. Generalversammlung deutscher Bergarbeiter Verband, 9 April 1898, Dortmund
Protokoll der 10. Generalversammlung deutscher Bergarbeiter Verband, 1 April 1899
Protokoll der 11. Generalversammlung deutscher Bergarbeiter Verband, 14–16 April 1900, Altenburg
Protokoll der 12. Generalversammlung deutscher Bergarbeiter Verband, 26–28 May 1901, Kassel
Protokoll der 13. Generalversammlung deutscher Bergarbeiter Verband, 17–19 May 1902, Essen
Protokoll der 14. Generalversammlung deutscher Bergarbeiter Verband, 1–3 June 1903, Zwickau
Protokoll der 17. Generalversammlung deutscher Bergarbeiter Verband, 5–9 May 1907, Dortmund
Protokoll der 18. Generalversammlung deutscher Bergarbeiter Verband, 16–21 May 1909, Eisenach
Protokoll der 19. Generalversammlung deutscher Bergarbeiter Verband, 21–26 May 1911, Bochum
Protokoll der 20. Generalversammlung deutscher Bergarbeiter Verband, 27 April to May 1912, Hannover
Protokoll der 21. Generalversammlung deutscher Bergarbeiter Verband, 15–21 July 1919, Bielefeld
Protokoll der 22. Generalversammlung deutscher Bergarbeiter Verband, 24–26 January 1920, Bochum
Protokoll der 23. Generalversammlung deutscher Bergarbeiter Verband, 29 May to 4 June 1921, Gleßen
Protokoll der 24. Generalversammlung deutscher Bergarbeiter Verband, 15–19 June 1924, Dresden
Protokoll der 25. Generalversammlung deutscher Bergarbeiter Verband, 4–8 July 1926, Saarbrücken
Geschäftsbericht Verband der Bergarbeiter Deutschlands 1904
Geschäftsbericht Verband der Bergarbeiter Deutschlands, 1905–06
Geschäftsbericht Verband der Bergarbeiter Deutschlands, 1907–08
Geschäftsbericht Verband der Bergarbeiter Deutschlands, 1909–10
Geschäftsbericht Verband der Bergarbeiter Deutschlands, 1911–12
Geschäftsbericht Verband der Bergarbeiter Deutschlands, 1913–14
Geschäftsbericht Verband der Bergarbeiter Deutschlands, 1915–16
Geschäftsbericht Verband der Bergarbeiter Deutschlands, 1917–18
Material zur Lage der Bergarbeiter während des Weltkrieges. Eine Sammlung von Eingabe der vier Bergarbeiterverbände, besonders des Vorstandes des Verbandes der Bergarbeiter Deutschlands, an die Regierungen sowie Zivil-, Militärbehörden Werkbesitzer, usw. Herausgegeben vom Vorstand des Verbandes der Bergarbeiter Deutschlands, Bochum, 1919.

*Gewerkverein christlicher Bergarbeiter Deutschlands*
Geschäftsbericht des Vorstandes des Gewerkvereins christlicher Bergarbeiter Deutschland für Jahre 1905 und 1906 und Protokoll der 11. Generalversammlung 29 June–1 July 1907, Gelsenkirchen
Geschäftsbericht des Vorstandes des Gewerkvereins christlicher Bergarbeiter Deutschland für Jahre 1907 und 1908 und Protokoll der 12. Generalversammlung 27–30 June 1909, Saarbrücken
Geschäftsbericht des Vorstandes des Gewerkvereins christlicher Bergarbeiter Deutschland für Jahre 1909 und 1910 und Protokoll der 13. Generalversammlung 9–12 July 1907
Bericht des Hauptvorstandes des Gewerkvereins christlicher Bergarbeiter Deutschlands, 1911 und 1912
Protokoll der 14. Generalversammlung 6–9 July 1913, Aachen
Bericht des Hauptvorstandes des Gewerkvereins christlicher Bergarbeiter Deutschlands, 1913 to 1918
Bericht über das 25jährige Jubelfest und Protokoll der 15. Generalversammlung des Gewerkvereins christlicher Bergarbeiter Deutschlands, 24–27 August 1919, Essen
Bericht des Hauptvorstandes des Gewerkvereins christlicher Bergarbeiter Deutschlands, 1921 to 1925
Protokoll der 18. Generalversammlung 13–15 May 1913, Saarbrücken
Bericht der Sozialisierungskommission über die Frage der Sozialisierung des Kohlenbergbaues, Berlin, Hans Robert Engelmann, 1920
Protokoll über die Verhandlungen des Bergarbeiter-Delegiertentagen für Preußen (einberufen von er Siebener Kommission), 28–30 March 1905, Berlin
Protokoll des 2. Bergarbeiter-Delegiertentagen für Preußen (einberufen von er Siebener Kommission), 11–12 February 1906, Berlin
Verhandlung der Sozialisierungskommission über den Kohlenbergbau im Jahre 1918/1919, Berlin 1921
Protokoll vom 1. Reichsbetriebsrätekongress für den Bergbau, Verband des Bergarbeiter Deutschlands, 6–7 November 1921, Magdeburg

**2. Hauptstaatsarchiv Düsseldorf**
*Regierung Düsseldorf/Arnsberg:*
Police: Monthly and Activity Reports 1920–26 (STAD 26119–26125)
Präsidialbüro (STAD 583, 588, 596)

*Political Files:*
Political Associations (STAD 15817–18)
Communist Party and Works' councils (STAD 15689, 15690, 15693, 16910, 16922, 16932, 16934, 16940, 16961)
Council Organisations (STAD 15279)
Unrest March to April 1921 (STAD 15822)
Rathenau Murder (STAD 16893)
Miners' strikes (STAD 15032)
Miners' unrest (STAD 16868)
Strikes 1922–1927 (STAD 16863–6)
Unemployment (STAD 16845, 16847, 16850)
Social-democratic and anarchist movement (STAD 9028, 9039–44, 9046, 42809)
Miners' strike 1905 (STAM 9056)

Social-democratic women's movement 1905–1913 (STAD 42789)
Confessional workers' associations and Christian trade unions 1901–1913 (STAD 15921)
National Workers' Associations 1910–1913 (STAD 15923)
Wiarus Polskie (STAD 42815)

**3. Staatsarchiv Münster**
*Oberbergamt:*
Wages (STAM OBA 1390, 1392, 1396–7)
Works Safety (STAM OBA 372, 373, 1781, 1778, 1792, 1793, 1795, 1811, 1813, 1815, 1849, 1850–5, 1862, 1864, 1866)
Co-operatives (STAM OBA 1799, 1800)
Schnappskasinos (STAM OBA 1834)
1905 Strike (STAM OBA 1843–48)
1912 Strike (STAM OBA 1857–9)
Foreign Labour during World War 1 (STAM OBA 1159–60)
Knappschaft (STAM OBA 258–61)

*Regierung Münster:*
Syndicalists (STAM 14442)
Polish Movement (STAM 14044–45, 14064, 14063)
Evangelical Associations (STAM 14391–93)
Revolutionary Movements (STAM 14427)
Communist Movement (STAM 14447–9, 14452, 14454, 14455)
Ruhr Occupation (STAM 14515, 14520)
Works Council Elections (STAM 14481)
Büro Kolpin (STAM 18, 36, 104, 128, 159, 179, 192, 197, 210, 215, 327, 337–40, 342–3, 355)

**4. Stadtarchiv Bochum**
*Landrat Bochum:*
SPD (STAB LA 1303–9, 1323)
KPD, USPD (STAB LA 1314–5)
Miners' movement (STAB 1269, 1274, 1282, 1284–5)
Alter Verband (STAB B314–5, 317)

**5. Bergbau Museen, Bochum (Bochum Mining Museum)**
*Records of the Zechenverband/Verein für bergbauliche Interessen:*
Collective Agreements and Wage negotiations 1919–1933 (BBA 90/4–91/4)
Newsletters 1911–1927 (BBA 3/14–7/14)
Reports 1924–1932 (BBA 235/15–1083/15)
Working time 1921–1940, 1919–1923, 1924–1940, (BBA 180/15,194/15, 197/15)
Working time propaganda 1919–1939 (BBA 181/15)
Working time in the mining industry 1921–1928 (BBA 186/15)
Working time in 'hot' places 1919–1940 (BBA 189/15)
Production in German and foreign, private and state mines 1919–1940 (BBA 231/15)
Works Council Law 1919–1922 (BBA 1085/15)
Social Questions 1920–1944 (BBA 296/15)
Food Supply 1923–1944 (BBA 302/15)
Reichknappschaft Committees 1921–1944 (BBA 320/15)

Market and Economic Conditions 1920–1940 (BBA 428/15)
Coal Prices 1919–1935 (BBA 437/15)
Central Coal Council 1919–1938 (BBA 505/15)
Negotiations and Correspondence 1919–1930 (BBA 131/15)
Administration 1919–1944 (BBA 854/15)
Improvement in mining 1906–1938 (BBA 569)
Mechanisation 1925–1926, 1926–1940 (BBA 625, 626)
Coal Cutters 1905–1938 (BBA 634)
Conveyors 1907–1943, 1901–1943 (BBA 817, 818)
Horses 1909–1944 (BBA 821)

*State mines:*
Collective Agreements July 1922–November 1925 (BBA 442/72)
Wage Statistics November 1924–February 1935 (BBA 445/72)
Labour Issues November 1918–August 1922 (BBA 512/72)
Works Councils August 1922–December 1925 (BBA 513/72)
Strikes 1924–1925, 1926–1931 (BBA 517/72, 518/72)
Housing August 1919–December 1925 (BBA 607/72)
Reparations July 1924–February (BBA 676/72)
Activity of the Vereins für bergbauliche Interessen July 1924–February 1926 (BBA 1031/72)

*Krupp mines:*
Strikes 1907–1925 (BBA 563/20)
Works History 1914–1923 (BBA 565/20)
Workers' Committee 1918–1919, 1919–1921 (BBA 992/20, 993/20)
Works' Councils 1919–1920, 1920–1923 (BBA 994/20, 995/20)

*Shamrock and Hibernia mines:*
Strikes at Shamrock 1912–1918 (BBA 42731/32)
Strikes at Hibernia November 1918–March 1919, April 1919–January 1922 (BBA 4283/32, 4284/32)
Agitation at Hibernia (BBA 4288/32)
Strikes at Hibernia March 1920–1926 (BBA 4290/32)

**6. Institut für Zeitungsforschung, Dortmund**
Centre Party Handbills and Leaflets (IZF Mappe 24)
KPD Handbills and Leaflets (IZF Mappe 10)
SPD Handbills and Leaflets (IZF Mappe 11)

**7. Essen City Library**
Die Arbeiterbewegung im Ruhrgebiet. Eine Gabe an den Parteitag 1907 vom Sozialdemokratischen Verein Kreis Essen, 1907.

## C. Newspapers and journals
*Aberdare Leader*
*Amman Valley Chronicle and East Carmarthen News*
*Bergarbeiter Zeitung*
*Bergknappe*

*Carmarthen Journal*
*The Celtic News and West Wales General Advertiser*
*Colliery Guardian*
*Colliery Workers' Magazine*
*Cymmer Searchlight*
*Deutsche Arbeit*
*Dortmunder Zeitung*
*Emscher Zeitung*
*Essener Allgemeine Zeitung*
*Essener Arbeiter Zeitung*
*Essener Volkszeitung*
*Free Press of Monmouthshire*
*Gelsenkirchener Allgemeine Zeitung*
*Glamorgan County Times*
*Glamorgan Free Press and Rhondda Leader*
*The Glamorgan Gazette: Bridgend and Neath Chronicle and Central Glamorgan Gazette, Incorporated*
*Y Geninen*
*Glückauf. Berg- und Hüttenmännische Zeitschrift*
*The Labour Voice*
*Märkischer Sprecher*
*Merthyr Express*
*Merthyr Pioneer*
*Monmouthshire Evening Post*
*Monmouthshire Merlin and South Wales Advertiser*
*Neath Gazette and Mid-Glamorgan Herald*
*Die Neue Zeit*
*Pontypool Free Press*
*Pontypridd Chronicle*
*Pontypridd Observer*
*Porth Gazette*
*Red Dawn*
*Rheinisch-Westfälische Arbeiter Zeitung*
*Rheinisch-Westfälische Zeitung*
*The Rhondda Leader*
*Rhondda Socialist*
*Ruhr Echo*
*Seren Gomer Press*
*Sozialistische Monatshefte*
*South Wales Argus*
*South Wales Daily News*
*South Wales Daily Star*
*South Wales Daily Telegraph*
*South Wales Worker*
*Tarian y Gweithiwr*
*Tremonia*
*Y Traethodydd*
*Western Mail*
*Westfälischer Kämpfer*

*Westfälische Volkszeitung*
*Welsh Outlook*
*The Workers' Dreadnought*
*Volksblatt*

## D. Contemporary works and pamphlets

Alter Verband, *Die Kämpfe um Knappschafts-Reformen im Allgemein Knappschaftsverein Bochum* (Bochum: 1910).

Barwich, F., *Der kommunistische Aufbau des Syndikalismus im Gegensatz zum Partei-Kommunismus und Staatssozialismus* (Berlin: Der Syndikalist, 1920).

Bezirkskommission der Sozialdemokratischen Partei für das Westliche Westfalen, *Der Essener Meineidsprozeß gegen Schröder im Wiederaufnahmeverfahren* (Dortmund: A Gerisch, 1911).

Bredenbeck, A., *1889: Die Erste Erhebung der Bergarbeiter. Zur Erinnerung an den Grossen Bergarbeiterstreik vor 20 Jahren* (Dortmund: A. Gerisch, 1909).

Brust, A., 'Der Bergarbeiterstreik im Ruhrrevier', *Archiv für Sozialwissenschaft und Sozialpolitik* (Tübingen: J. C. B. Monr, 1905), 480–506.

Christlicher Arbeiterbewegung, *Sozialdemokratischer Wahlschwindel vor Gericht. Ein Beweis für die skrupellose Bekämpfung des Gewerkvereins Christlicher Bergarbeiter* (Cologne: Chritlicher Gewerschaftsverlag, 1910).

Cook, A. J., *Is it Peace?* (London: Workers' Publications Ltd., 1926).

Engel, K., *Zum Ausstande der Bergarbeiter im Ruhrbezirke* (Berlin: Julius Springer, 1905).

Effert, J., *Zur Geschichte der Bergarbeiterbewegung im Ruhrrevier* (Essen: 1906).

Erdmann, A., *Die Christlichen Arbeiterbewegung in Deutschland* (Stuttgart: J. H. W. Dietz, 1908).

Evans, D., *Labour Strife in the South Wales Coalfield 1910–1911* (Cardiff: Educational Publishing Co. Ltd., 1911).

Freie Vereinigung deutsches Gewerkschaften, *Das Programm des freien Vereinigung deutsches Gewerkschaften* (Berlin, 1906).

Harries, F. J., *History of Conservatism in the Rhondda* (Pontypridd: Glamorgan County Times Office, 1912).

Hartshorn, V., *Mr. Baldwin Attacks Miners' Hours and Wages* (London: The Labour Party, 1926).

Hauptvorstand des Gewerkvereins christlicher Bergarbeiter Deutschlands, *Denkschrift über die Ursachen der Arbeits-Einstellungen in Ruhrkohlenbergbau* (Essen, 1917).

Hay, W. F. and N. Ablett, 'A minimum wage for miners: What it means, and how to get it', in Thomas Mann, *Miners, Wake Up* (The Industrial Syndicalist), 1911.

Hay, W. F., *Industrial Democracy for Miners: A Plan for the Democratic Control of the Mining Industry* (Porth: D. A. Davies, 1919).

Hay, W. F., *Education and the Working Class* (Liverpool: Liverpool and District Council for Independent Working-class Education, 1920).

Hilferding, R., *Die Sozialisierung und die Machtverhältnisse der Klassen* (Berlin: Verlagsgenossenshcaft Freiheit, 1920).

Hodges, F., *Workers' Control in the Coal Mining Industry* (London: Mines for the Nation Campaign Committee, 1920).

Hodges, F., *Nationalisation of the Mines* (London: Leonard Parsons, 1920).

Horner, A., *One Mineworkers' Union – Why?* (London: National Minority Movement, 1926).

Horner, A. and N. Watkins, *British Mineworkers' Union: A Constitution* (London, 1926).

Hue, O., *Neutrale oder parteiische Gewerkschaft. Ein Beitrag zur Gewerkschaftsfrage, zugleich eine Geschichte der deutschen Bergarbeiterbewegung* (Bochum: Möller, 1900).

Hue, O., *Mehr Bergarbeiterschutz* (Bochum: Möller, 1900).

Hue, O., *Unsere Taktik beim Generalstreik* (Bochum: Hansmann & Co., 1905).

Hue, O., *Die Bergarbeiter. Historische Darstellung der Bergarbeiter-Verhältnisse von der Ältesten bis in die neueste Zeit* (Berlin: J. H. W. Dietz, 1913/1981).

Hue, O., *Die Sechsstundenschicht im Bergbau* (Bochum: Vorstand des Verbandes der Bergarbeiter Deutschlands, 1920).

Hue, O., *Die Sozialisierung der Kohlenwirtschaft* (Berlin: Vorwärts, 1921).

Imbusch, H., *Ist eine Verschmelzung der Bergarbeiter-Organisationen möglich?* (Essen: Fredebeul, 1906).

Imbusch, H., *Arbeitsverhältnis und Arbeiter-Organisationen im deutschen Bergbau. Eine geschichtliche Darstellung* (Berlin/Bonn: J. H. W. Dietz, 1908/1980).

Imbusch, H., *Bergarbeiter-Streik im Ruhrgebiet in Frühjahr 1912* (Cologne: Christl. Gewerkschaftsverlag, 1912).

Imbusch, H., *Die Gelben in der deutschen Arbeiterbewegungen* (Cologne: Christliche Gewerekschaftsverlag, 1912).

Imbusch, H., *Die Tätigkeit und Erfolge des Gewerkvereins christlicher Bergarbeiter Deutschlands*, Verlag des Gewerkvereins christlicher Bergarbeiter Deutschlands (Essen: Verlag des Gewerkvereins christlicher Bergarbeiter Deutschlands, 1915).

Imbush, H., *25 Jahre Gewerkverein christlicher Bergarbeiter* (Essen: Verlag des Gewerkvereins christlicher Bergarbeiter Deutschlands, 1919).

Imbusch, H., *August Brust. Ein Lebensbild des Gründers der ersten christlichen Gewerkschaft* (Berlin: Christlicher Gewerkschafts-Verlag, 1924).

Industrial Committee of the South Wales Socialist Society, 'A Plan for the Democratic Control of the Mining Industry', 1919.

Jevons, H. S., *The British Coal Trade* (London: Kegan Paul, 1915).

Jones, T. I., *Mining Royalties and All About Them* (London: ILP, 1908).

Liberal Publication Department, *The Coal Mining Industry: Recent History, Causes of Unrest and The Liberal Remedy* (London, 1925).

Lensing, L., *Der Große Bergarbeiter-Streik des Jahres 1889 im Rheinisch-Westfälischen Kohlenrevier. Ein Wort zur Abwehr* (Dortmund, 1889).

Levenstein, A., *Die Arbeiterfrage mit besonderer Berücksichtigung der Sozialpsychologischen Seite des modernen Grossbetriebes und der psycho-physischen Einwirkungen auf die Arbeiter* (Munich: Ernst Rheinhardt, 1912).

Lütgenau, Dr F., *Was wollen die Syndikalisten?* (Berlin: Verlag für praktische Politik und geistige Erneuerung, 1920).

Müller, O., *Die christliche Gewerkschaftsbewegung Deutschlands mit besonderer Berücksichtigung der Bergarbeiter- und Textilarbeiter-Orgnisationen* (Karlsruhe: G Braunschen Hofbuchdruckerei, 1905).

Natorp, G., *Der Ausstand der Bergarbeiter im niederrheinisch-westfälischen Industriebezirk* (Essen: G. D. Bädeker, 1889).

Neunerkommission für die Vorbereitung der Sozialisierung des Bergbaues im rheinisch-westfälischen Industriegebiet, *Die Sozialisierung des Bergbaues und der Generalstreik im rheinisch-westfälischen Industrigebiet* (Essen, 1919).

Oldenburg, Dr K., *Studien zur Rheinisch-Westfälischen Bergarbeiterbewegung* (Leipzig: Duncker and Humbolt, 1890).

Osterroth, N., *Otto Hue. Ein Lebensbild für seine Freunde* (Bochum: H. Hausmann, 1922).

Osthold, P., *Die Geschichte des Zechenverbandes 1908–1913. Ein Beitrag Zur Deutschen

*Sozialgeschichte* (Berlin: Elsner, 1934).

Pieper, L., *Die Lage der Bergarbeiter im Ruhrrevier* (Stuttgart and Berlin: J. G. Cotta'sche, 1903).

Price, E., *The History of Penuel: Calvanistic Methodist Church* (Wrexham: Hughes and Son, 1925).

Redmayne, R., *The British Coal-Mining Industry during the War* (Oxford: Clarendon Press, 1923).

Schriften der Gesellschaft fur Soziale Reform, *Aufsätze bei den Streik der Bergarbeiter im Ruhrgebiet* (Jena: Gustav Fischer, 1905).

Seippel, M., *Bochum einst und jetzt. Ein Rück- und Rundblick Bei der Wende des Jahrhunderts* (Bochum: Schürmann und Klagges, 1901).

Severn, M., *The Miners' Evangel. A Text Book for All Manual Workers* (Pontypridd: John W. Ford, 1895).

Spethman, H., *Zwölf Jahre Ruhrbergbau. Aus seiner Geschichte von Kriegsanfang bis zum Franzosenabmarchen 1914–1925. Band 1: Aufstand und Ausstand vor und nach dem Kapp-Putsch bis zur Ruhrbesetzung* (Berlin: Reimar Hobbing 1929).

Spethmann, H., *Zwölf Jahre Ruhrbergbau. Aus seiner Geschichte von Kriegsanfang bis zum Franzosenabmarchen 1914–1925. Band 2: Aufstand und Ausstand vor und nach dem Kapp-Putsch bis zur Ruhrbesetzung* (Berlin: Reimar Hobbing, 1929).

Spethmann, H., *Zwölf Jahre Ruhrbergbau. Aus seiner Geschichte von Kriegsanfang bis zum Franzosenabmarchen 1914–1925. Band 3: Der Ruhrkampf 1923 bis 1925 in seinen Leitlinien* (Berlin: Reimar Hobbing, 1929).

Spethmann, H., *Zwölf Jahre Ruhrbergbau. Aus seiner Geschichte von Kriegsanfang bis zum Franzosenabmarchen 1914–1925. Band 4, Der Ruhrkampf 1923 bis 1925. Das Ringen um die Kohle* (Berlin: Reimar Hobbing, 1930).

Spethmann, H., *Die Rote Armee an Ruhr und Rhein. Aus Kapptagen* (Berlin: Reimar Hobbing, 1930).

Spethmann, H., *Der Maistreik 1924 im Ruhrbergbau. Ein grundsätzlicher Arbeitskampf* (Berlin: Reimar Hobbing, 1932).

Thomas, D. A., *Some Notes on the Present State of the Coal Trade in the United Kingdom with Special Reference to that of South Wales and Monmouthshire Together with a Proposal for the Prevention of Undue Competition and for Maintaining Prices at a Remunerative Level*, (Cardiff, 1896).

Unofficial Reform Committee, *The Miners' Next Step. Being a Suggested Scheme for the Reorganisation of the Federation*, ed. R. Merfyn Jones (London: Pluto Press, 1973 (first published 1912)).

Vincent, J. E., *John Nixon: Pioneer of the Steam Coal Trade in South Wales* (London: John Murray, 1900).

Vogelsang, H., *Verfassung und Verwaltung des Gewerkvereins christlicher Bergarbeiter Deutschlands* (Cologne: Christlicher Gewerkschafts-Verlag, 1910).

Vogelsang, H., *Geschichte, Verfassung und Verwaltung des Gewerkvereins christlicher Bergarbeiter Deutschlands* (Essen: Verlag des Gewerkvereins christlicher Bergarbeiter Deutschlands, 1915).

Vogelsang, H., *Der Weltkrieg: Deutschlands Bergarbeiter im Kriege* (Essen: Sekretariat Sozialer Studentenarbeit, 1917).

Vogelsang, H., *Reichsarbeitsgemeinschaft für den Bergbau* (Essen: Gewerkverein christlicher Bergarbeiter Deutschlands, 1920).

Volksverein für das katholische Deutschlands, *Bergarbeiterschutz und Zentrum. Eine aktenmäßige Darstellung der Tätigket des Zentrums im Deutschen Reichstage und im*

*Preußischen Landtage zugunsten der Bergarbeiter* (M.Gladbach: Volksverein-Verlag Gmbh., 1908).

Vorstand des Verbandes der Bergarbeiter Deutschlands, *Die Streikjustiz im Ruhrrevier oder Madame Justitia in 'verbundenen' Augen* (Bochum: H. Hansmann, 1913).

Vorstand des Verband der Bergarebtier Deutschlands, *Material zur Lage der Bergarbeiter während des Weltkrieges. Eine Sammlung von Eingabe der vier Bergarbeiterverbände, besonders des Vorstandes des Verbandes der Bergarbeiter Deutschlands, an die Regoerungen sowie Zivil-, Militärbehörden, Werksbesitzer, usw.* (Bochum, 1919).

Vorstand des Verbandes der Bergarbeiter Deutschlands, *Unternehmer und Kommunisten während der Bergarbeiterkämpfe im Mai 1924* (Bochum: Hansmann & Co., 1924).

Werner, G., *Was kann der deutsche Bergbau von England lernen?* (Essen: Fredebeul & Koenen, 1910).

Werner, G., *Wie die Wetter schlagen. Ein Kulturbild aus dem Ruhrrevier* (Essen: Deutschen Steigerverbandes, 1913).

Werner, G., *Unfälle und Erkrankungen im Ruhr-Bergbau* (Essen: Fredebeul, 1920).

Werner, G., *Müssen die Wetter schlagen?* (Berlin: Verlag des Verfasser, 1921).

## E. Autobiographies, memoirs and novels

Coombes, B. L., *These Poor Hands: The Autobiography of a Miner Working in South Wales* (London: Gollancz, 1939).

Davies, W. H., *The Right Place, the Right Time: Memories of Boyhood Days in a Welsh Mining Community* (Swansea: Christopher Davies, 1975).

Davies, W. H., *Ups and Downs* (Swansea: Christopher Davies, 1975).

Edwards, W. J., *From the Valley I Came* (London: Angus and Robertson, 1956).

Faust, O., *Vom Bremsjungen zum Betriebesinspektor. Ein Leben im Ruhrbergbau (1867–1914)* (Essen: Klartext, 1989).

Griffiths, J., *Pages from Memory* (London: J. M. Dent & Sons, 1969).

Hodges, F., *My Adventures as a Labour Leader* (London: George Newnes Ltd, 1925).

Horner, A., *Incorrigible Rebel* (London: MacGibbon & Kee, 1960).

Jones, B. and C. Williams (eds), *With Dust Still in his Throat: A B. L. Coombes Anthology* (Cardiff: University of Wales Press, 1999).

Marchwitza, H., *Meine Jugend* (Berlin: Tribüne, 1952).

Morgan, Revd J. V., *The War and Wales* (London: Chapman and Hall Ltd, 1916).

Stonelake, E., *The Autobiography of Edmund Stonelake*, ed. Anthony Mor-O'Brien (Aberdare: Mid-Glamorgan Education Committee, 1981).

Paynter, W., *My Generation* (London: Allen & Unwin, 1972).

Severing, C., *1919 – 1920: Im Wetter und Watterwinkel. Aufzeichnungen und Erinnerungen des Saatsministers A.D. Carl Severing Ehemaligen Reichs und Staatsminister in Befehlsbereich des VII Armeekorps* (Bielefeld: Greven, 1927).

Sommerlad, P., *Vom 'Hessebub' Zum 'Bochumer Jungen': 70 Jahre Sozialistische Bewegung in Bochum (1878–1948)*, ed. Peter Friedemann and Edgar Heinevetter (Essen: Klartext, 1990).

Stenbock-Fermor, Graf A., *Meine Erlebnisse als Bergarbeiter* (Stuttgart: J Engelhorns, 1928).

Stewart, W., *J. Keir Hardie: A Biography* (London: ILP, 1921).

Werner, G., *Ein Kumpel. Erzählung aus dem Leben der Bergarbeiter* (Berlin, 1929).

## F. Reference works

*Deutsche Biographische Enzyklopäde*
*Dictionary of Labour Biography*
*Digest of Welsh Historical Statistics*
*The Labour Who's Who 1927*
*Rheinisch-Westfälische Wirtschaftsbiographien*
*Sozialdemokratische Parlamentarier in den deutschen Reichs- und Landtagen, 1867–1933*

## G. Selected secondary literature

The extensive historiographies of both south Wales and the Ruhr coalfields coupled with the constraints of space make it impossible to present an exhaustive bibliography. Readers can find fuller references in the endnotes or may consult my PhD thesis 'Political Identities: The Miners of the Ruhr and South Wales, 1890–1926' (University of Glamorgan, 2004).

Abrams, L., *Workers' Culture in Imperial Germany: Leisure and Recreation in the Rhineland and Westphalia* (London and New York: Routledge, 1992).

Adelman, G., *Die soziale Betriebsverfassung des Ruhrbergbaus vom Anfang des 19. Jahrhunderts bis zum Ersten Weltkrieg unter besonderer Berücksichtigung des Industrie- und Handelskammerbezirks Essen* (Bonn: Ludwig Röhrscheid Verlag, 1962).

Anderson, M. L., *Practicing Democracy: Elections and Political Culture in Imperial Germany* (Princeton: Princeton University Press, 2000).

Arnot, R. P., *South Wales Miners, Glowyr de Cymru: A History of the South Wales Miners' Federation, 1898–1914*, vol. 1 (London: Allen & Unwin, 1967).

Arnot, R. P., *South Wales Miners, Glowyr de Cymru: A History of the South Wales Miners' Federation, 1914–1926*, vol. 2 (Cardiff: Cymric Federation Press, 1975).

Baber, C. and L. J. Williams (eds), *Modern South Wales: Essays in Economic History* (Cardiff: University of Wales Press, 1986).

Beddoe, D., *Out of the Shadows: A History of Women in Twentieth-Century Wales* (Cardiff: University of Wales Press, 2000).

Belchem, J. and N. Kirk (eds), *Languages of Labour* (Aldershot: Ashgate, 1997).

Berg, W., *Wirtschaft und Gesellschaft in Deutschland und Großbritannien im Übergang zum 'Oganisierten Kapitalismus'. Unternehmer, Angestellte, Arbeiter und Staat im Steinkohlenbergbau des Ruhrgebietes und von Südwales, 1850–1914* (Berlin: Duncker & Humbolt, 1984).

Berg, W., 'Zwei Typen industriegesellschaftlicher Modernisierung: Die Bergarbeiter im Ruhrgebiet und in Südwales im 19. und frühen 20. Jahrhundert', in Gustav Schmidt (ed.), *Bergbau in Großbritannien und im Ruhrgebiet: Studien zur vergleichenden Geschichte des Bergbaus, 1850–1930* (Bochum: Brockmeyer, 1985).

Berger, S., *The British Labour Party and the German Social Democrats, 1900–1931* (Oxford: Clarendon Press, 1994).

Berger, S., 'Working-Class culture and the labour movement in the south Wales and the Ruhr coalfields: A Comparison', *Llafur*, 8:2 (2001), 5–40.

Berger S., A. Croll and N. LaPorte (eds), *Towards a Comparative History of Coalfield Societies* (Aldershot: Ashgate, 2005).

Biagini, E. F. and A. J. Reid (eds), *Currents of Radicalism. Popular Radicalism, Organised Labour and Party Politics in Britain, 1850–1914* (Cambridge: Cambridge University Press, 1991).

Blackbourn, D. and G. Eley (eds), *The Peculiarities of German History: Bourgeois Society and Politics in Nineteenth-Century Germany* (Oxford: Oxford University Press, 1984).

Bock, H. M., *Geschichte des linken Radikalismus in Deutschland. Ein Versuch* (Frankfurt am Main: Suhrkamp, 1976).

Bock, H. M., 'Anarchosyndicalism in the German labour movement: A rediscovered minority tradition', in Marcel von der Linden and Wyne Thorpe (eds) *Revolutionary Syndicalism: An International Perspective* (Aldershot: Scolar Press, 1990), pp. 59–80.

Boyns, T., 'Work and death in the south Wales coalfield, 1874–1914', *Welsh History Review* 12: 4 (1985), 514–37.

Brephol, W., *Der Aufbau des Ruhrvolkes im Zuge der Ost-West-Wanderung* (Recklinghausen: Bitter and Co., 1948).

Breuilly, J., *Labour and Liberalism in Nineteenth-Century Europe: Essays in Comparative History* (Manchester: Manchester University Press, 1992).

Brüggemeier, F.-J., 'Soziale Vagabundage oder revolutionär Heros? Zur Sozialgeschichte Der Ruhrbergarbeiter 1880–1920', in Lutz Niethammer (ed.) *Lebenserfahrung Und Kollektive Gedächtnis. Die Praxis Der >>Oral History<<* (Frankfurt am Main: Syndikat, 1980), pp. 193–213.

Brüggemeier, F.-J., *Leben Vor Ort. Ruhrbergleute und Ruhrbergbau, 1889–1919* (Munich: Beck, 1984).

Budde, H., *Christliche-Sozial an der Ruhr. Eine Volksbewegung im Zentrum der Industrie. Dokumentation Eines 150jährigen Wegen* (Cologne: Orthen Druck, 1988).

Bulmer, M. I., 'Sociological models of the mining community', *Sociological Review* 23:1 (1975), 61–92.

Burghardt, U., *Die Mechanisierung des Ruhrbergbaus 1890–1930* (Munich: Beck, 1995).

Campbell, A., N. Fishman and D. Howell (eds), *Miners, Unions and Politics, 1910–47* (Aldershot: Scolar, 1996).

Childers, T., 'The social language of politics in Germany: The sociology of political discourse in the Weimar Republic', *American Historical Review* 95:2 (1990), 331–58.

Church, R., *The History of the British Coal Industry, 1830–1913: Victorian Pre-Eminence*, vol. 3 (Oxford: Clarendon Press, 1986).

Church, R. and Q. Outram (eds), *Strikes and Solidarity: Coalfield Conflict in Britain, 1889–1966* (Cambridge: Cambridge University Press, 1998).

Claggett, W., J. Loesch, W. P. Shively and R. Snell, 'Political leadership and the development of political cleavages: Imperial Germany, 1871–1912', *American Journal of Political Science* 26 (1982), 643–64.

Crew, D., *Town in the Ruhr: A Social History of Bochum, 1860–1914* (New York: Colombia University Press, 1979).

Crew, D., 'Regionale Mobilität und Arbeiterklasse. Das Beispiel Bochum, 1880–1901' in K. Schönhoven and Dieter Langwiesche (eds), *Arbeiter in Deutschland: Studien zur Lebensweise der Arbeiterschaft im Zeitalter der Industrialisierung* (Paderborn: Ferdinand Schöningh, 1981), pp. 81–97.

Croll, A., '"People's remembrancers" in a post-modern age: Contemplating the non-crisis of Welsh labour history', *Llafur* 8:1 (2000), 5–17.

Croon, H. and K. Utermann (eds), *Zeche und Gemeinde: Untersuchungen über den Strukturwandel einer Zechengemeinde im nördlichen Ruhrgebiet* (Tübingen: J.C.B. Moheer (Paul Siebeck), 1958).

Daunton, M. J., 'Down the pit: Work in the Great Northern and south Wales coalfields, 1870–1914', *Economic History Review* 34:4 (1981), 578–97.

David, W., 'The Labour Party and the "exclusion" of the Communists: The case of the Ogmore Divisional Labour Party in the 1920s', *Llafur* 3:4 (1983), 5–15.

Davies, K., 'Rival Prophets? Williams Ferris Hay, Noah Ablett and the debate over working-class political action in the south Wales coalfield 1910–1914', *Llafur* 7:3/7:4 (1998/99), 89–100.

Davies, P., *A. J. Cook* (Manchester: Manchester University Press, 1987).

Ditt, K. and D. Kift (eds), *1889: Bergarbeiterstreik Und Wilhelminische Gesellschaft* (Hagen: Westfälisches Industriemuseum, 1989).

Düwell, K. and W. Köllmann (eds), *Rheinland-Westfalen im Industriezeitalter*, 3 volumes (Wuppertal: Hammer, 1984).

Edwards, N., *History of the South Wales Miners' Federation* (London: Lawrence and Wishart, 1938).

Egan, D., 'The Unofficial Reform Committee and the Miners' Next Step: Documents from the W. H. Mainwaring papers with an introduction and notes', *Llafur* 2:3 (1978), 64–80.

Egan, D., 'Noah Ablett 1883–1935', *Llafur* 4:3 (1986), 19–30.

Eisenberg, C., 'The comparative view in labour history: Old and new interpretations of the English and German labour movements before 1914', *International Review of Social History* 34 (1989), 403–30.

Eley, G., 'Nations, publics, and political cultures: Placing Habermas in the nineteenth century', in Craig Calhoun (ed.), *Habermas and the Public Sphere* (Cambridge, Mass and London: MIT Press, 1992), pp. 289–339.

Evans, E. W., *Mabon: William Abraham 1842–1922: A Study in Trade Union Leadership* (Cardiff: University of Wales Press, 1959).

Evans, E. W., *The Miners of South Wales* (Cardiff: University of Wales Press, 1961).

Evans, N. and K. Sullivan, 'Yn llawn o dân cymreig (Full of Welsh Fire): The language of politics in Wales 1880–1914', in Geraint H. Jenkins (ed.), *A History of the Welsh Language: The Welsh Language and Its Social Domains 1801–1911* (Cardiff: University of Wales Press, 2000), pp. 561–85.

Evans, R. J. (ed.), *Society and Politics in Wilhelmine Germany* (London: Croon Helm, 1978).

Fagge, R., *Power, Culture and Conflict in the Coalfields: West Virginia and South Wales 1900–1912* (Manchester: Manchester University Press, 1996).

Faulenbach, B. and G. Högel (eds), *Eine Partei in Ihrer Region. Zur Geschichte der SPD im westlichen Westfalen* (Essen: Klartext, 1988).

Feldman, G. D., *Army, Industry and Labour in Germany 1914–1918* (New Jersey: Princeton University Press, 1966).

Fischer, C., *The Ruhr Crisis, 1923–1924* (Oxford: Oxford University Press, 2003).

Fisk, M. J., *Housing in the Rhondda, 1800–1940* (Cardiff: Merton Priory Press Ltd, 1996).

Foot, M., *Aneurin Bevan: A Biography 1897–1945*, vol. 1 (London: MacGibbon & Kee, 1962).

Francis, H., 'The anthracite strike and the disturbances of 1925', *Llafur* 1:2 (1973), 53–66.

Francis, H. and D. Smith, *The Fed: A History of the South Wales Miners in the Twentieth Century* (Cardiff: University of Wales Press, 1998).

Fricke, D., *Der Ruhrbergarbeiterstreik von 1905* (Berlin: Rütten & Loening, 1955).

Friedemann, P. and G. Seebold (eds), *Struktueller Wandel und Kulturelles Leben: Politische Kultur in Bochum 1860–1990* (Essen: Klartext, 1992).

Gehrmann, S., *Fußball, Vereine, Politik. Zur Sportgeschichte des Reviers 1900–1940* (Essen: Reimar Hobbing, 1988).

Gilbert, D., *Class, Community, and Collective Action: Social Change in Two British Coalfelds, 1850–1926* (Oxford: Clarendon Press, 1992).

Goch, S., *Arbeiterbewegung und Arbeiterkultur im Ruhrgebiet. Eine Untersuchung am Beispiel Gelsenkirchen, 1848–1975* (Düsseldorf: Dorste Verlag, 1990).

Goch, S. and L. Heidemann (eds), *100 Jahre Bismarck. Ein Stadtteil 'mit besondersm Erneuerungsbedarf* (Essen: Klartext, 2001).

Gorlas, J. and D. J. K Peukert (eds), *Ruhrkampf 1920* (Essen: Klartext, 1986).

Grant, R., *The Parliamentary History of Glamorgan, 1542–1976* (Swansea: Christopher Davies, 1978).

Gregory, R., *The Miners and British Politics, 1906–1914* (Oxford: Oxford University Press, 1968).

Griffiths, R., *S. O. Davies: A Socialist Faith* (Llandysul: Gomer Press, 1983).

Guttsman, W. L., *The German Social Democratic Party, 1875–1933: From Ghetto to Government* (London: George Allen & Unwin, 1984).

Habermas, J., *The Theory of Communicative Action: Reason and the Rationalization of Society*, trans. Thomas McCarthy (Cambridge: Polity Press, 1984).

Habermas, J., *The Theory of Communicative Action: The Critique of Functionalist Reason*, trans. Thomas McCarthy (Cambridge: Polity Press, 1987).

Haffert, C. D., *Die katholischen Arbeitervereine Westdeutschlands der Weimarer Republik* (Essen: Klartext, 1994).

Hartewig, K., *Das unbrechenbare Jahrzehnt: Bergarbeiter und ihre Familien im Ruhrgebiet 1914–1924* (Munich: Beck, 1993).

Hartmann, K., *Der Weg zur gewerkschaftliche Organisation. Bergarbeiterbewegung und kapitalischer Bergbau im Ruhrgebiet, 1851–1889* (Munich: tuduv-Verlagsgesellschaft, 1977).

Heid, L. and J. H. Schoeps (eds), *Arbeit und Alltag im Revier. Arbeiterbewegung und Arbeiterkultur im Westlichen Ruhrgebiet im Kaiserreich und in der Weimarer Republik* (Duisburg: Walter Braun Verlag, 1985).

Hennicke, O., *Die Rote Ruhrarmee* (Berlin: Ministry for National Defence, 1956).

Henze, J., *Sechstundenschicht im Ruhrbergbau 1918–1920: Ursachen und Verlauf eines Arbeitszeitkonflikts* (Freiburg: Burg-Verlag, 1988).

Herbert, T. and G. E. Jones (eds), *Wales Between the Wars* (Cardiff: University of Wales Press, 1988).

Hickey, S., *Workers in Imperial Germany: The Miners of the Ruhr* (Oxford: Clarendon Press, 1985).

Hiepel, C., *Arbeiterkatholizismus an der Ruhr: August Brust und der Gewerkverein christlicher Bergarbeiter* (Stuttgart: Kohlhammer, 1999).

Hoffman, W., 'Politik in der Provinz: Kommunale Politisierung dargestellt an den sechs Industriedörfen im Landkreis Recklinghausen vom Beginn der Industrialisierung bis 1914', *Bochumer Historische Studien, Neuere Geschichte* 16 (1996).

Holtfrerich, C.-L., *Quantitative Wirtschaftsgeschichte des Ruhrkohlenbergbaus im 19. Jahrhunderts. Eine Führungssektoranalyse* (Dortmund: Ardey-Verlag, 1973).

Hopkin, D. R. and G. S. Kealey (eds), *Class, Community and the Labour Movement: Wales and Canada, 1850–1930* (Aberystwyth: Llafur/CCLH, 1985).

Howard, C., 'Reactionary Radicalism: The Mid-Glamorgan By-Election, March 1910', *Glamorgan Historian* 9 (1973), 29–41.

Howard, C., '"Expectations Born to Death": The local Labour Party expansion in the 1920s', in Jay M. Winter (ed.), *The Working Class in Modern British History: Essays in Honour of Henry Pelling* (Cambridge: Cambridge University Press, 1983), pp. 65–81.

Howell, D., *British Workers and the Independent Labour Party, 1888–1906* (Manchester: Manchester University Press, 1983).

Hughes, G. A. (ed.), *Men of No Property: Studies of Welsh Trade Unionism* (Caerwys: Gwasg Gwenffrwd, 1971).

Jackson, J. H., *Migration and Urbanization in the Ruhr Valley* (New Jersey: Humanities Press, 1997).

Jäger, W., *Bergarbeitermilieus und Parteien im Ruhrgebiet. Zum Wahlverhalten des katholischen Bergarbeitermilieus bis 1933* (Munich: Beck, 1996).

James L. S., 'War and Industry: A study of the industrial relations of the mining regions of South Wales and the Ruhr during the Great War, 1914–1918', *LHR*, 68:2 (2003), 195–215

Jantke, C., *Bergmann und Zeche. Die sozialen Arbeitsverhältnisse einer Schachtanlage des nördlichen Ruhrgebiets in der Sicht der Bergbau* (Tübingen: J. C. B. Mohr (Paul Siebeck), 1953).

Joerss, R., 'Moralisch Vorkommen und politisch Gefährlich? Die Schnapskasinos im Ruhrgebiet am Ende des 19. Jahrhunderts', *Beiträge zur Geschichte Dortmunds und der Grafschaft Mark*, 92/93 (2001/2002), 133–74.

John, A. V., 'A miner struggle? Women's protests in Welsh mining history', *Llafur* 4:1 (1984), 72–90.

John, A. V. (ed.), *Our Mothers' Land: Chapters in Welsh Women's History, 1830–1939* (Cardiff: University of Wales Press, 1991).

Jones, G. S., *Languages of Class: Studies in English Working Class History, 1832–1982* (Cambridge: Cambridge University Press, 1983).

Jones, J. G., 'Wales and the "New Socialism", 1926–29', *Welsh History Review* 11:2 (1982), 173–99.

Jones, P. N., *Colliery Settlement in the South Wales Coalfield, 1850–1926* (Hull: University of Hull, 1969).

Joyce, P., *Visions of the People: Industrial England and the Question of Class, 1848–1914* (Cambridge: Cambridge University Press, 1991).

Kerr, C. and A. Siegel, 'The interdisciplinary propensity to strike – An international comparison' in Robert Dubin, Arthur Kornhauser, Arthur M. Ross (eds), *Patterns of Industrial Conflict* (New York,: McGraw Hill, 1954), pp. 189–212.

Kift, D. (ed.), *Kirmes – Kneipe – Kino. Arbeiterkultur im Ruhrgebiet zwischen Kommerz und Kontrolle (1850–1914)* (Paderborn: Ferdinand Schöningh, 1992).

Kirchhoff, H. G., *Die staatliche Sozialpolitik im Ruhrbergbau 1871–1914* (Cologne: Westdeutscher Verlag, 1957).

Klan, U. and D. Nelles, *'Es lebt noch eine Flamme'. Rheinische Anarcho-Syndikalisten/innen in der Weimarer Republik und im Faschismus* (Grafenau-Döffingen: Trotzdem-Verlag, 1986).

Kleßmann, C., *Polnische Bergarbeiter im Ruhrgebiet 1870–1945* (Göttingen: Vandenhoeck & Ruprecht, 1978).

Kluge, U., 'Essener Sozialisierungsbewegung und Volkswehr Bewegung im Rheinisch-Westfälischen Industriegebiet 1918/19', *Internationale Wissenschaftliche Korrespondenz* 16 (1972), 56–65.

Koch, M. J., *Die Bergarbeiterbewegung im Ruhrgebiet zur Zeit Wilhelms II (1889–1914)* (Düsseldorf: Dorste-Verlag, 1954).

Korff, G., 'Volkskultur und Arbeiterkultur. Überlegungen am Beispiel der Sozialistischen Maifestradition', *Geschichte und Gesellschaft*, 5:1 (1979), 83–102.

Kotowski, G., 'Auf dem Boden der gegebenen vollendeten Tatsachen! Der politische Katholizismus', in Detlef Lehnert and K. Megerle (eds), *Politische Identität und Nationale Gedenktage. Zur Politischen Kultur in der Weimarer Republik* (Opladen:

Westdeutscher Verlag, 1989), pp. 157–80.

Kreuzer, C., *Mit Kreuz und Hammer. Katholische Arbeiterschaft in Bochum und Hattingen 1863 bis 1933. Ein Beitrag zur Geschichte der Arbeiterbewegung* (Bochum: Schürmann & Klagges, 1979).

Kroker, E., 'Arbeiterausschüsse im Ruhrgebiet zwischen 1906 und 1914', *Der Anschnitt* 6:30 (1978), 204–15.

Kroker, E. and W. Kroker, *Solidarität aus Tradition. Die Knappenvereine im Ruhrgebiet* (Munich: Beck, 1988).

Kühr, H., *Parteien und Wahlen im Stadt- und Landkreis Essen in der Zeit der Weimarer Republik* (Düsseldorf: Droste Verlag, 1973).

Kulczycki, J. J., *The Foreign Worker and the German Labour Movement: Xenophobia and Solidarity in the Coal Fields of the Ruhr, 1871–1914* (Oxford: Berg, 1994).

Lawrence, J., *Speaking for the People: Party, Language and Popular Politics in England, 1867–1914* (Cambridge: Cambridge University Press, 1998).

Lehnert, D., '"Staatspartei der Republik" oder "Revolutionaere Reformisten"? Die Sozialdemokraten', in Detlef Lehnert and K. Megerle (eds), *Politische Identität und nationale Gedenktage. Zur Politischen Kultur in der Weimarer Republik* (Opladen: Westdeutscher Verlag, 1989), pp. 89–113.

Lepsius, R., 'Parteiensystem und Sozialstruktur. Zum Problem der Demokratisierung der deutschen Gesellschaft', in Gerhard A. Ritter (ed.), *Die deutschen Parteien vor 1918* (Cologne: Kiepenheuer & Witsch, 1973), 56–80.

Lewis, E. D., *The Rhondda Valleys* (Cardiff: University College Cardiff Press, 1958).

Lewis, R., *Leaders and Teachers: Adult Education and the Challenge of Labour in South Wales, 1906–1940* (Cardiff: University of Wales, 1993).

Lidtke, V. L., *The Alternative Culture: Socialist Labour in Imperial Germany* (Oxford: Oxford University Press, 1985).

Lieven, M., *Senghennydd: The Universal Pit Village, 1890–1930* (Llandysul: Gomer Press, 1994).

Lucas, E., *Märzrevolution 1920. Vom Generalstreik gegen den Militärputsch zum bewaffneten Arbeiteraufstand*, vol. 1 (Frankfurt: Roter Stern, 1970).

Lucas, E., *Märzrevolution 1920. Der bewaffnete Arbeiteraufstand im Ruhrgebiet in seiner inneren Struktur und in seinem Verhältnis zu den Klassenkämpfen in den verschiedenen Regionen des Reiches*, vol. 2 (Frankfurt am Main: Roter Stern, 1973).

Lucas, E., *Märzrevolution 1920. Verhandlungsversuche und deren Scheitern Gegenstrategien von Regierung und Militär; Die Niederlage der Aufstandsbewegung; Der Weiße Terror*, vol. 3 (Frankfurt am Main: Roter Stern, 1978).

Lucas, E., *Zwei Formen von Radikalismus in der deutschen Arbeiterbewegung* (Frankfurt am Main: Roter Stern, 1976).

Lützenkirchen, R., *Der sozialdemokratische Verein für den Reichstagswahlkkreis Dortmund-Hörde. Ein Beitrage zur Parteigeschichte* (Dortmund: Historischen Vereins Dortmund, 1970).

Macintyre, S., *Little Moscow: Communism and Working-Class Militancy in Inter-War Britain* (London: Croom Helm, 1980).

Maier, C. S., *Recasting Bourgeois Europe: Stabilization in France, Germany and Italy in the Decade after World War 1* (Princeton: Princeton University Press, 1988).

Mattheier, K. J., *Die Gelben. Nationale Arbeiter zwischen Wirtschaftsfrieden und Streik* (Düsseldorf: Schwann, 1973).

Matthews, I., 'The world of the anthracite miner', *Llafur*, 6:1. (1992), 96–104.

May, E., 'Labour, Capital and the State in the South Wales Coalfield 1912–1921: The Case of Gwaun-cae-Gurwen', in Ian Blanchard (ed.), *New Directions in Economic and Social History* (Avonbridge: Newless Press, 1995), pp. 47–55.

McKibbin, R., *The Evolution of the Labour Party, 1910–1924* (Oxford: Oxford University Press, 1974).

McKibbin, R., *The Ideologies of Class: Social Relations in Britain 1880–1950* (Oxford: Clarendon Press, 1990).

Minchinton, W. E. (ed.), *Industrial South Wales 1750–1914: Essays in Welsh Economic History* (London: Frank Cass and Company Ltd, 1969).

Mommsen, H., 'Sozialpolitik im Ruhrbergbau', in Dietmar Petzina, Hans Mommsen, Bernd Weisbrod and Dirk Stegmann (eds), *Industrielle System und politische Entwicklung in der Weimarer Republik* (Düsseldorf: Dorste Verlag, 1974), pp. 300–21.

Mommsen, H. and Ulrich Borsdorf (eds), *Glück Auf, Kameraden: Die Bergarbeiter und ihre Organisationen in Deutschland* (Cologne: Bund-Verlag, 1979).

Mommsen, H. (ed.), *Arbeiterbewegung und industrieller Wandel. Studien zu gewerkschaftlichen Organisationsproblemen im Reich und an der Ruhr* (Wuppertal: Hammer, 1980).

Mommsen, W. J. and H.-G. Husung (eds), *The Development of Trade Unionism in Great Britain and Germany 1880–1914* (London: George Allen and Unwin, 1985).

Morgan, J., 'Police and Labour in the age of Lindsay, 1910–1936', *Llafur* 5:1 (1988), 14–20.

Morgan, K. O., 'D. A. Thomas: The industrialist as politician', *Glamorgan Historian* 3 (1966), 33–51.

Morgan, K. O., 'The New Liberalism and the challenge of Labour: The Welsh experience, 1885–1929', *Welsh History Review* 6:3 (1973), 288–312.

Morgan, K. O., *Keir Hardie: Radical and Socialist* (London: Weidenfeld and Nicolson, 1975).

Morgan, K. O., *Wales in British Politics 1868–1922*, 3rd edn, (Cardiff: University of Wales Press, 1980).

Morgan, K. O., *Rebirth of a Nation: A History of Modern Wales* (Oxford: Oxford University Press, 1981).

Morgan, P., 'Keeping the Legends Alive', in Tony Curtis (ed.), *Wales: The Imagined Nation: Studies in Cultural and National Identity* (Bridgend: Poetry Welsh Press, 1986), pp. 17–42.

Morgan, P., 'The gwerin of wales: Myth and reality', in I. Hume and W. T. R. Pryce (eds), *The Welsh and Their Country* (Llandysul: Gomer Press, 1986), pp. 134–52.

Mor-O'Brien, A., 'Patriotism on trial: The strike of the south Wales miners, July 1915', *Welsh History Review* 12:1 (1984), 76–104.

Mor-O'Brien, A., 'The Merthyr Boroughs election, November 1915', *Welsh History Review* 12:4 (1985), 538–66.

Morris, J. H. and L. J. Williams, *The South Wales Coal Industry, 1841–1875* (Cardiff: University of Wales Press, 1958).

Murphy, R. C., 'The Polish trade union in the Ruhr coal field: Labor organization and ethnicity in Wilhelmine Germany', *Central European History* 9:4 (1978), 335–47.

Neumann, W., *Die Gewerkschaften im Ruhrgebiet. Voraussetzung, Entwicklung und Wirksamkeit* (Cologne: Bund-Verlag, 1951).

Oenning, R. K., '"*Du da mitte polnische Farben ...*" *Sozialisationerfahrungen von Polen im Ruhrgebiet 1918 bis 1939*' (Münster: Waxmann, 1991).

Oertzen, P. von, 'Die grossen Streiks der Ruhrbergarbeiterschaft im Frühjahr 1919. Ein Beitrag zur Diskussion über die revolutinäre Entstehungsphase der Weimarer

Republik', *Vierteljahrshefte für Zeitgeschichte* 6:3 (1958), 231–62.

Oertzen, P. von, *Betriebsräte in der Novemberrevolution. Eine Politikwissenschaftliche Untersuchung über Ideengehalt und Struktur der betrieblichen und wirtschaftlichen Arbeiterräte in der deutschen Revolution 1918/19* (Berlin/Bonn-Bad Godersberg: J. H. W. Nachf., 1976).

O'Leary, P., *Immigration and Integration: The Irish in Wales, 1798–1922* (Cardiff: University of Wales Press, 2000).

Parisius, B., *Lebenswege im Revier. Erlebnisse und Erfahrungen zwischen Jahrhundertwwende und Kohlenkrise – erzählt von Frauen und Männern aus Borbeck* (Essen: Rainer Henselocsky, 1984).

Parry, J., 'Trade unionists and early socialism in south Wales, 1890–1908', *Llafur* 4:3 (1985), 43–54.

Parry, J., 'Labour leaders and local politics 1888–1902: The example of Aberdare', *Welsh History Review* 14:3 (1989), 399–416.

Patch, W. L., *Christian Trade Unions in the Weimar Republic, 1918–1933. The Failure of 'Corporate Pluralism'* (New Haven and London: Yale University Press, 1985).

Paul, J., *Alfred Krupp und die Arbeiterbewegung* (Düsseldorf: Schwam, 1987).

Pelling, H., *Social Geography of British Elections, 1885–1900* (London: Macmillan, 1967).

Peukert, D., 'Industrialisierung des Bewußtseins? Arbeitserfahrugen von Ruhrbergleuten im 20. Jahrhundert', in K. Tenfelde (ed.), *Arbeit und Arbeitserfahrung in der Geschichte* (Göttingen: Vandanhoeck and Ruprecht, 1986), pp. 92–119.

Pitt, R., 'Educator and agitator: Charlie Gibbons 1888–1967', *Llafur* 5:2 (1989), 72–83.

Pope, R., *Building Jerusalem: Nonconformity, Labour and the Social Question in Wales, 1906–1939* (Cardiff: University of Wales Press, 1998).

Pope, R., 'Facing the dawn: Socialists, nonconformists and Llais Llafur, 1906–1914', *Llafur* 7:3/7:4 (1998/99), 77–87.

Pretty, D. A., 'David Morgan (Dai O'r Nant), miners' Agent. A portrait of leadership in the south Wales coalfield', *Welsh History Review* 20:3 (2000), 495–531.

Przigoda, S., *Unternehmenverbände im Ruhrbergbau. Zur Gesichichte von Bergbau-Verein und Zechenverband 1858–1933* (Bochum: Deutschen Bergbau-Museums, 2002).

Reid, D., 'Reflections on labor history and language', in Lenard R. Berlanstein (ed.), *Rethinking Labor History: Essays on Discourse and Class Analysis* (Chicago: University of Illinois Press, 1993), pp. 39–54.

Reulecke, J. (ed.), *Arbeiterbewegung an Rhein und Ruhr* (Wuppertal: Hammer, 1974).

Reulecke, J. and W. Weber (eds), *Fabrik, Familie, Feierabend. Beiträge zur Sozialgeschichte des Alltags im Industriezeitalter* (Wuppertal: Hammer, 1978).

Rimlinger, G. V., 'International differences in the strike propensity of coal miners: Experience in four countries', *Industrial and Labour Revolutions Review* (1959), 389–405.

Ritter, G. A., *Die Arbeiterbewegung im Wilhelminischen Reich. Die Sozialdemokratische Partei und die freien Gewerkschaften 1890–1900* (Berlin-Dahlem: Colloqium Verlag, 1963).

Ritter, G. A., *Arbeiterbewegung, Parteien und Parlamentarismus. Aufätze zur Deutschen Sozial- und Verfassungsgeschichte des 19. und 20. Jahrhunderts* (Göttingen: Vandenhoeck & Ruprecht, 1976).

Ritter, G. A. (ed.), *Der Aufsteig der deutschen Arbeiterbewegung. Sozialdemokratie und freie Gewerkschaften im Parteiensystem und Sozialmilieu des Kaiserreichs* (Munich: R. Oldenburg, 1990).

Rohe, K., *Vom Alten Revier zum heutigen Ruhrgebiet. Die Entwicklung einer regionalen politischen Gesellschaft im Spiegel der Wahlen*' (Königstein/Ts: Anton Hain Meisenheim, 1979).

Rohe, K., 'Politische Kultur und ihre Analyse. Probleme und Perspektiven der Politische Kulturforschung', *Historische Zeitschrift* 250 (1990), 321–46.

Rosenblum, N. J. and Post, R. C. (eds), *Civil Society and Government* (Princeton: Princeton University Press, 2002).

Rübner, H., *Freiheit und Brot. Die Freie Arbeiter-Union Deutschlands. Eine Studie zur Geschichte des Anachosyndikalismus* (Cologne: Libertad Verlag, 1994).

Ruck, M., *Die Freien Gewerkschaften im Ruhrkampf 1923* (Cologne: Bund-Verlag, 1986).

Rürup, R. (ed.), *Arbeiter und Soldatenräte im Rheinisch-Westfälischen Industriegebiet* (Wuppertal: Hammer, 1975).

Salaman, G., 'Some sociological determinants of occupational communities', *Sociological Review* 19:1 (1971), 53–77.

Savage, M., *The Dynamics of Working-Class Politics: The Labour Movement in Preston, 1880–1940* (Cambridge: Cambridge University Press, 1987).

Schäfer, M., *Heinrich Imbusch: Christlicher Gewerkschaftsführer und Widerstandskämpfer* (Munich: Beck, 1990).

Schneider, M., *Die christlichen Gewerkschaften 1894–1933* (Bonn: Verlag Neue Gesellschaft, 1982).

Schneider, M., 'Religion and labour organization: The Christian trade unions in the Wilhelmine Empire', *European Studies Review* 12:3 (1982), 345–69.

Schönhoven, K., *Expansion und Konzentration. Studien zur Entwicklung der Freien Gewerkschaften im Wilhelminischen Deutschland 1890 bis 1914* (Stuttgart: Klett-Cotta, 1980).

Smith, D. (ed.), *A People and a Proletariat: Essays in the History of Wales 1780–1980* (London: Pluto, 1980).

Smith, D., 'Tonypandy 1910: Definitions of community', *Past and Present* 87 (1980), 158–84.

Smith, D., *Wales! Wales?* (London: Allen & Unwin, 1984).

Smith, D., 'From riots to revolt: Tonypandy and the Miners' Next Step', in Trevor Herbert and Gareth Elwyn Jones (eds), *Wales 1880–1914* (Cardiff: University of Wales Press, 1988), pp. 107–38.

Smith, D., *Aneurin Bevan and the World of South Wales* (Cardiff: University of Wales Press, 1993).

Spencer, E. G., 'Employer response to unionism: Ruhr coal industrialists before 1914', *Journal of Modern History* 3, 48:3 (1976), 397–412.

Spencer, E. G., *Management and Labor in Imperial Germany: Ruhr Industrialists as Employers, 1896–1914* (New Jersey: Rutgers University Press, 1984).

Sperber, J., 'The shaping of political Catholicism in the Ruhr basin, 1848–1881', *Central European History* 16:4 (1983), 347–67.

Stead, P., 'Vernon Hartshorn: miners' agent and cabinet minister', *Glamorgan Historian* 6 (1970), 83–94.

Stead, P., 'Establishing a heartland: The Labour Party in Wales', in Kenneth D. Brown (ed.), *The First Labour Party* (Beckenham: Croom Helm, 1985), pp. 64–88.

Steinisch, I. and K. Tenfelde, 'Technischer Wandel und soziale Anpassung in der Deutschen Schwerindustrie während des 19. und 20. Jahrhunderts', *Archiv für Sozialgeschichte*, 28 (1988), 27–74.

Supple, B., *The History of the British Coal Industry 1913–1946: The Political Economy of Decline*, vol. 4 (Oxford: Clarendon Press, 1987).

Suval, S., *Electoral Politics in Wilhelmine Germany* (London and Chapel Hill: University of North Carolina Press, 1985).

Tampke, J., 'The rise and fall of the Essen model, January-February 1919', *Internationale Wissenschaftliche Korrespondenz* 13:2 (1977), 160–72.

Tampke, J., *The Ruhr and Revolution. The Revolutionary Movement in the Rhenish-Westphalian Industrial Region 1912–1919* (London: Croom Helm, 1979).

Tanner, D., *Political Change and the Labour Party, 1900–18* (Cambridge: Cambridge University Press, 1990).

Tanner, D., C. Williams and D. R. Hopkin (eds), *The Labour Party in Wales 1900–2000* (Cardiff: University of Wales Press, 2000).

Tenfelde, K., 'Die "Krawalle von Herne" im Jahre 1899', *Internationale Wissenschaftliche Korrespondenz* (1979), 71–104.

Tenfelde, K., 'Der bergämmische Arbeitsplatz während der Hochindustrialisierung (1890 to 1914)', in Werner Conze and Ulrich Engelhardt (eds), *Arbeiter im Industrialisierungspross. Herkunft, Lage und Verhalten* (Stuttgart: Ernst Klett, 1979), pp. 283–335.

Tenfelde, K., 'Bergarbeiterkultur in Deutschland. Ein Überblick', *Geschichte und Gesellschaft* 5:1 (1979), 12–53.

Tenfelde, K., *Sozialgeschichte der Bergarbeiterschaft an der Ruhr im 19. Jahrhundert*, 2nd edn (Bonn: Verlag Neue Gesellschaft, 1981).

Tenfelde, K., 'On the History of Industrial Relations in Mining', in Gerald D. Feldman and K. Tenfelde (eds), *Workers, Owners and Politics in Coal Mining: An International Comparison of Industrial Relations* (Oxford: Berg, 1990), pp. 1–11.

Thomas, B., 'A cauldron of rebirth: Population and the Welsh language in the nineteenth century', in Geraint H. Jenkins (ed.), *A Social History of the Welsh Language: The Welsh Language and its Social Domains 1801–1911* (Cardiff: University of Wales Press, 2000), pp. 81–99.

Trentmann, F. (ed.), *Paradoxes of Civil Society: New Perspectives on Modern German and British History* (New York and Oxford: Berghahn Books, 2000).

Tschirbs, R., *Tarifpolitik im Ruhrbergbau 1918–1933* (Berlin and New York: Walter de Gruyter, 1986).

Wallraff, H. J., 'Die Belastung einer Gewerkschaft durch ideologische Differenzen – Spannungen innerhalb der christlichen Gewerkschaftsbewegung in den Jahren 1900–1914', in H. O. Vetter (ed.), *Vom Sozialistengesetz zur Mitbestimmung. Zum 100. Geburtstag von Hans Böckler* (Cologne: Bund Verlag, 1975), pp. 135–52.

Walters, R. H., *The Economic and Business History of the South Wales Coal Industry*, (New York: Arno Press, 1977).

Weisbrod, B., 'Entrepreneurial politics and industrial relations in mining in the Ruhr region: From managerial absolutism to co-determination', in G. D. Feldman and K. Tenfelde (eds), *Workers, Owners and Politics in Coal Mining: An International Comparison of Industrial Relations* (Oxford: Berg, 1990).

Weitz, E. D., *Creating German Communism, 1890–1990: From Popular Protests to Socialist State* (Princeton: Princeton University Press, 1997).

Welskopp, T., *Das Banner der Brüderlichkeit. Die deutsche Sozialdemokratie vom Vormärz bis zum Sozialistengesetz* (Bonn: Dietz, 2000).

White, J., 'Syndicalism in a mature industrial setting: The case of Britain', in Marcel von der Linden and Wyne Thorpe (eds), *Revolutionary Syndicalism: An International Perspective* (Aldershot: Scolar Press, 1990), pp. 101–18.

Williams, C., 'An able administrator of capitalism? The Labour Party in the Rhondda,

1917–1921', *Llafur* 4:4 (1987), 20–33.
Williams, C., *Democratic Rhondda: Politics and Society, 1885–1951* (Cardiff: University of Wales Press, 1996).
Williams, C., 'Democracy and nationalism in Wales: The Lib-Lab engima', in Robert Stradling, Scott Newton and David Bates, *Conflict and Coexistence. Nationalism and Democracy in Modern Europe: Essays in Honour of Harry Hearder* (Cardiff: University of Wales Press, 1997), pp. 107–31.
Williams, C., *Capitalism, Community and Conflict: The South Wales Coalfield, 1898–1947* (Cardiff: University of Wales Press, 1998).
Williams, C., 'The odyssey of Frank Hodges', *Transactions of the Honourable Society of Cymmrodorion*, 5 (1999), 110–30.
Williams, G. A., *When Was Wales? A History of the Welsh* (Harmondsworth: Penguin, 1985).
Williams, J., *Was Wales Industrialised? Essays in Modern Welsh History* (Llandysul: Gomer Press, 1995).
Williams, J., 'The Fed: Birth and first steps', *Llafur* 7:3/7:4 (1998/9), 67–75.
Williams, L. J., 'The new unionism in south Wales, 1889–92', *Welsh History Review* 1:4 (1963), 413–29.
Woodhouse, M. G., 'Mines for the nation or mines for the miners? Alternative perspectives on industrial democracy, 1921–1912', *Llafur* 2:3 (1978), 92–109.

## H. Unpublished dissertations

Davies, D. K., 'The Influence of Syndicalism and Industrial Unionism on the South Wales Coalfield, 1898–1921: A Study in Ideology and Practice' (PhD dissertation, University of Wales, 1991).
Gesau, K. F., 'Syndikalismus in der Ruhrbergarbeiterschaft zu Beginn der Weimarer Republik, 1918–1925' (Magisterarbeit, Münster, 1983).
McCook, B., 'Borders of Integration: Polish Migrant Workers in the Ruhr Valley of Germany and Pennsylvania Anthracite Regions of the United States, 1870–1924' (PhD dissertation, University of California, Berkeley, 2004).
May, E., 'A Question of Control: Social and Industrial Relations in the South Wales Coalfield and the Crisis of Post-War Reconstruction' (PhD dissertation, University of Wales, 1995).
Schwaderer, C., 'Die Reichstagswahlen und die Wahlen zur Nationalversammlung im Landkreis und der Stadt Bochum 1912–1920' (Hausarbeit der ersten Staatsprüfung für das Lehramt an Gymnasien, Bochum, 1980).
Tiedtke, H.-W., 'Die Reichstagswahlen und die Reichstagswahlkämpfe in südlichen Teil des Ruhrgebiets von der Gründung des Nord-Deutschen Bundes bis zum Beginn des Ersten Weltkrieges' (Hausarbeit der Fachprüfung für das Lehramt am Gymnasium, Bochum, 1976).

# Index